BUILDING & INSTALLING
ELECTRONIC INTRUSION ALARMS
THIRD EDITION

John Cunningham has had many years of experience in various phases of electronics. After five years of electronic servicing, he spent eight years in the instrument department of General Electric Company. Later as chief special products engineer at Canoga Electronics, Inc., he was responsible for design and development of missile range equipment including radar and telemetry systems. For many years he has been a consultant in the security electronics field, and he is Senior Technical Director at Cleveland Institute of Electronics.

Mr. Cunningham is a senior member of the Institute of Electrical and Electronics Engineers. Another SAMS book by Mr. Cunningham is *Cable Television* (Second Edition).

BUILDING & INSTALLING ELECTRONIC INTRUSION ALARMS
THIRD EDITION

By John E. Cunningham

International Standard Book Number: 0-672-21954-9
Library of Congress Catalog Card Number: 82-50021

Edited by: *Louis Keglovits*
Illustrated by: *Wm. D. Basham*

Printed in the United States of America.

PREFACE

This book is intended to allow anyone with a knowledge of elementary electronics to construct intrusion alarms that will fill the needs of almost any security requirement. To this end very sophisticated circuits have been avoided and circuits have been selected that use readily available components.

There will be a temptation for the reader to go directly to the chapter in the book which attracts his interest. The author strongly suggests that before doing this, or at least before starting the construction of any system, the reader carefully read Chapter 1. This chapter has been included in this edition to alert the reader to problems that might be encountered if he starts constructing one of the systems described in the other chapters, without adequate forethought. Much of the material in this chapter was obtained from correspondence and phone conversations with readers of earlier editions.

Chapter 2 describes some of the components used in following chapters that might be new to technicians or enthusiasts who have been concerned primarily with entertainment equipment such as tv and stereo sets. It is assumed that the reader is already familiar with the more common components used in the various circuits.

Chapter 3 is devoted to trigger circuits. The trigger circuit is the heart of any intrusion alarm. It is also responsible for most false alarms. Any of the systems described in following chapters should be used with a good trigger circuit.

Chapters 4 through 8 cover various types of alarm systems. The remaining chapters are devoted to various accessories used with intrusion alarms, and selecting a system for a particular application. Chapter 16 is included to help anyone who has problems in making one of the circuits in the book operate properly.

JOHN E. CUNNINGHAM

ACKNOWLEDGMENTS

So many people have helped in gathering the material used in this book that it is impossible to list them all. Particular mention should be made of my son Robert who helped construct and test the circuits, Steve Simcic of Cleveland Institute of Electronics who was most helpful with his suggestions for automobile protection, Steve Eldard of SRS systems, and Bonnie Cohen of LSI Computer Systems, Inc.

A debt of gratitude is due to the readers of previous editions of the book who have shared their experiences and made suggestions. Last, but not least, I am thankful to Grace Slavik who, in spite of a serious handicap, has always been an inspiration. Her persistent encouragement made the book possible.

CONTENTS

BEFORE YOU START

This chapter appears at the beginning of the book to help the reader avoid many of the problems that are frequently encountered in selecting or building an intrusion alarm. Most of these problems can be avoided by adequate planning before the project is started.

Much too often, an alarm is completely constructed and partially installed before the builder discovers that it will not work in a particular application. Of course, the most serious problem is when an alarm system does not provide adequate protection. Building such an alarm is a waste of time and money. Another problem almost as bad is a system which produces such a large number of false alarms that it cannot be trusted.

In this chapter we discuss many of the things that should be considered before an alarm project is started. By carefully taking all of these things into consideration, an alarm can be selected that will provide very good protection. Neglecting important considerations can lead to complete dissatisfaction.

TO BUILD OR NOT TO BUILD

After all of the factors covered in the remainder of the chapter are taken into consideration, the reader must decide whether to build an alarm system, or to buy a commercially available system. In some applications the only response that can be expected from an intrusion alarm would be from a private security patrol company. Such companies often furnish the alarm system as a part of their overall package. In such an application the best action may be to turn the entire problem over to the security company.

One good reason not to build a system is to save money. True, unless one counts the time required to build and install it, a home-built system will usually cost less than a commercially available alarm, but an alarm system is no place to try to save money. The purpose of the system is to protect something, and usually one of the major differences between a good protective system and a bad one is the amount of money spent on the system.

A good reason for buying a system would be if after reading the book, the reader did not feel at home with the system he selects to build. Many problems with home-built security systems can be traced to the builder's inability to understand and test the system.

In favor of building a system is the fact that the system can be tailored more closely to a given application than a commercial system that is designed for more universal applications. Furthermore, the home constructed system has the very important advantage that the builder is the only one who knows exactly how it works—unless he is foolish enough to explain it to someone else.

This point cannot be stressed too strongly. There is a great temptation for the builder of anything to show it off to his friends and acquaintances. There are many cases on record where security systems have been foiled because the intruder knew exactly how the system functioned.

In this respect, the builder should take a lesson from banks. Everyone knows that a bank has at least one protective system, but no one knows exactly how it works or where the wiring is located. The electrical plans of a bank building rarely show the wiring of the security system.

WHAT GOOD IS IT?

Many people make the mistake of expecting an alarm system to do more than it is capable of doing. An alarm system should be thought of as a system that will produce an alarm signal, either locally or at a remote location, whenever an intrusion takes place. This signal can be used in many ways:

1. A loud local alarm may scare a burglar away before he has a chance to steal anything or do any other harm.
2. It may alert someone who can take further action. This might be a security guard or a sleeping resident, or even a neighbor who will call the police.

3. It might summon the police, or a private security-guard service to the scene to apprehend the intruder.

The effectiveness of an intrusion alarm depends on the response time; that is, the elapsed time between when the alarm is tripped and when someone arrives on the scene. This is a very important concept, particularly if whatever is being protected has a high value. It isn't uncommon for prospective burglars to actually measure the response time of a system. This is accomplished by doing something that will trip the alarm and then hiding safely while waiting to see how much time elapses before someone arrives on the scene.

SELECTING THE SYSTEM

Assuming that the reader has evaluated the risk and has arranged for some of the other security provisions just described, the remaining task is to select the system. The advantages and limitations of the various systems are described in other chapters; however, there are some general considerations that will help in selecting a system.

The first step is to make what might be called an intrusion inventory. This is a list of the different ways in which an intruder might enter the protected area. Remember that if the expected reward is great enough, an intruder will resort to drastic methods of gaining entry to the premises.

Obvious ways of entering include:

1. Doors. The lock might be picked or the door forced open.
2. Windows. The window might be forced open, or the glass broken.
3. Other routes. When the property being protected is very valuable, or when an intruder might think it is valuable, any possible route might be taken. This includes ventilator shafts or even cutting through the roof or walls.

All routes that an intruder can be expected to take should be protected.

In selecting a system it should be remembered that the simpler the system, and the fewer components used, the more reliable the system is apt to be. In this respect, the electromechanical system is often the best. It almost always uses fewer parts. It is usually less susceptible to external influences such as temperature extremes and ambient noise than the other types. Of course, it has the disadvantage that separate detectors are required for each possible point of entry.

Other systems should usually be considered when the electromechanical system is difficult or impossible to tailor to a particular application. Sometimes a combination of two or more different types of systems will provide optimum protection.

WHERE WILL IT BE LOCATED?

Before actually constructing an intrusion alarm, some thought should be given as to where the circuitry will be located. It is usually a good idea to keep an intrusion alarm out of sight. The less people who know about it, the more effective it will be. This often leads to the system being located in some out-of the way place such as an attic. Unfortunately, such a place might be a really hostile environment for electronic equipment.

All electronic equipment is subject to the influence of ambient conditions such as temperature extremes, humidity, and ambient electrical noise. Of course, it is possible to design electronic equipment to operate properly in a hostile environment, but the best way to eliminate the influence of such hostile things is to avoid them as much as possible.

The circuits of an intrusion alarm should not be located where they will be subject to wide variations in temperature. Similarly, alarm circuits should not be located close to high current appliances that might cause electrical noise.

If extreme temperatures are unavoidable, the components used to build the system should be rated accordingly. This may be accomplished by selecting proper components and derating them in accordance with the manufacturer's instructions. If electrical noise cannot be avoided, the circuits must be well shielded and power supply lines should be well filtered. Again, the best approach is to avoid hostile environments.

IF YOU CHANGE THINGS

All of the circuits in this book have been built and tested. If the diagrams are followed carefully, and good components are used, the circuits should work properly. Unfortunately, few technicians ever follow a circuit exactly. Sometimes this is because the specified components are not available under some circumstances. In such cases it is necessary to make substitutions. Sometimes a circuit in a book such as this suggests something else to a reader. The circuit in the book is then merely a starting place and some additional design is done to arrive at the final circuit. There is little to be gained by simply telling the reader not to make any changes in the circuit. Many readers will make changes regardless of any caveat from the author. Instead, we will try to establish some guidelines that will help if the reader does decide to make changes either in components or in the circuit itself.

First, let's talk about components. Most technicians will try to use components that they happen to have on hand before they go to the expense of buying new ones. This is certainly understandable. It is also often a cause of problems. Probably the first consideration in using a component that happens to be available is to be sure that it is good. This means testing it.

Problems are often encountered when the capacitance of an electrolytic capacitor has changed or when the leakage current has increased. In many circuits one can use many different types of transistors without any ill effects. In other cases an available transistor will work marginally. This can lead to many annoying false alarms.

The ultimate test of any component is to actually try it in the circuit. It is strongly suggested that in cases like this the circuit be breadboarded and tested before final construction is started. It is much easier to make changes on a breadboard than on a final assembly.

As far as changes to the circuit itself are concerned, the situation isn't as clear. It isn't uncommon for readers to develop circuits for particular applications that are as good or better than those conceived by an author. On the other hand, many readers develop circuits that are not as good as those in a book.

Sometimes a good circuit can be developed mostly by chance, but this is very rare. Usually a circuit that is developed will be commensurate with the reader's knowledge and ability. Thus if you are familiar with electronic circuits and their design you may be successful at modifying the circuits in this book. If, on the other hand, you have little knowledge of the details of circuit design, you will do better to follow the diagrams in the book as closely as possible.

No claim is made that the circuits described in this book are the best possible for any particular purpose. They have been built and work well. If readers develop better circuits for any of the applications, the author would be happy to hear about them.

ALARM SYSTEM COMPONENTS

CHAPTER 2 ——————————————————————————

Some of the components used in alarm circuits may be unfamiliar to the reader because they are not commonly used in home-entertainment electronic equipment. Such components include the silicon controlled rectifier, or SCR, and special, integrated circuits, and pulse transformers. These components and their features are described in this chapter.

The technician should become thoroughly familiar with the principle of operation of each component used in a circuit that he intends to build. Once the basic principles are mastered, it is usually easy to make the circuit work.

THE SILICON CONTROLLED RECTIFIER

A very desirable feature of any intrusion alarm is that once the alarm is tripped, it should not be possible to shut it off without some special resetting arrangement. A component that lends itself quite well to such applications is the silicon controlled rectifier, which is also called a thyristor, or simply an SCR. The SCR operates like a switch that can be closed electrically and remains closed even after the switching current is removed.

Fig. 2-1 shows a diagram of an SCR. It has three terminals, the anode and cathode, which are similar to the corresponding elements in an ordinary diode, and a third electrode called a gate. The gate controls whether or not current will pass between the anode and the cathode.

The operating principle of the SCR can be understood from Fig. 2-2, which shows a circuit that can actually be used in an elementary intrusion alarm. The load, which may be a bell or a relay that operates

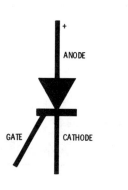

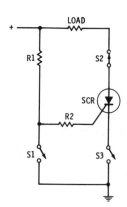

Figure 2-1.
Silicon controlled rectifier.

Figure 2-2.
Typical SCR circuit.

a bell, is connected between the positive side of the power supply and the anode of the SCR. The SCR acts as a switch between the bottom of the load and ground. When no positive voltage is applied to the gate, the SCR will appear as an open circuit, and no current will pass through the load.

In the circuit shown in Fig. 2-2, switch S1 is the protective switch and is normally closed, holding the gate at ground potential. The switch might be installed so that it will open if a door is opened. When this happens, a positive voltage is applied to the gate through resistors R1 and R2. This positive voltage will cause the SCR to turn on, thus connecting the load to ground. If the load were a bell, the bell would sound whenever switch S1 was opened.

One of the attractive features of the SCR for this application is that once it is turned on, it will not turn off even after switch S1 is closed. There is nothing that we can do at the gate to turn off the SCR once it has "fired." It will remain on as long as the current through it is greater than some value of current, usually called the *holding* or *keep-alive* current. The SCR will only turn off if the current can be reduced below this value. This may be accomplished in the circuit in Fig. 2-2 by interrupting the current by briefly opening either the anode or the cathode lead by opening switch S2 or S3, respectively. Either of these switches can be used to reset the circuit. Both are not necessary. When S3 is used, some of the anode current might try to pass through the gate circuit to ground. This current is limited to a safe value by resistor R2, which is usually in the order of a few hundred ohms.

An SCR will turn on very quickly. The gate need be made positive with respect to the cathode for only about 5 microseconds (μs) to turn the SCR on. Because of the charge stored in the device, it takes a little longer to turn it off. The current must be reduced below the value of the holding current for about 50 microseconds to be sure that the SCR will turn off. When a switch is used for interrupting the circuit, this is no problem. No switch that is operated by hand will act in 50 microseconds. If, however, an electronic circuit is used to turn off the SCR, the turn-off time must be considered.

There is another way that an SCR can be turned on. This is not intentional, but is a result of the way that the device is constructed. If the voltage between the anode and cathode should increase rapidly—faster than about 20 volts per microsecond (V/μs)—the SCR may turn on even though there is no source of positive voltage connected to the gate. This is called the dV/dT effect; dV represents the change in voltage occurring in the time interval dT.

Because of the dV/dT effect, which is inevitable in practical SCRs, the SCR may turn on when power is applied. This means that when an alarm circuit is first energized, the SCR may be in the on state causing the alarm to sound. This is annoying, but not as annoying as the fact that the alarm might also sound when power is restored after a power failure. This may well occur at a time when there is no one around to reset it.

Another very troublesome result of the dV/dT effect is that sharp noise spikes on the power lines can turn on the SCR. This would lead to an alarm circuit that would be very unreliable.

The dV/dT effect can be minimized by connecting a *snubber* circuit between the anode and cathode, as shown in Fig. 2-3. Here, the voltage between the anode and cathode cannot change too rapidly because the voltage across a capacitor cannot change instantaneously. The resistor is used to prevent the capacitor discharge current from damaging the SCR.

The snubber circuits shown in the alarm circuits in this book have been found to work with commonly available SCRs. If the

Figure 2-3. ——————————————
SCR "snubber" circuit.

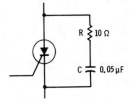

R ⩾ 10 Ω

C ⩵ 0.05 μF

builder finds that his SCR will turn on whenever power is applied to the circuit, he can usually correct the difficulty by experimenting with the snubber circuit.

SCR Specifications

The characteristics of an SCR can be defined in terms of a few specifications. By paying proper attention to the specifications, an experimenter can select an SCR that will adapt a circuit to his particular need. Some of the more important specifications are described below.

Peak Reverse Voltage (V_{RM})—The V_{RM} is the maximum voltage that can be applied to the SCR in the reverse direction; that is, with the anode negative and the cathode positive. If this voltage is exceeded, the SCR may break down and be destroyed.

Gate Trigger Current (I_{GT})—The I_{GT} is the current required at the gate to turn on the SCR. If the rated gate-trigger current of an SCR is too high for a particular circuit, there may not be enough energy available to turn on the SCR.

Peak Gate Voltage Forward (V_{GFM}) and Reverse (V_{GRM})—The symbols V_{GFM} and V_{GRM} represent the maximum forward and reverse voltages that can be applied between the gate and cathode under any conditions. Some circuits apply transient voltages to the gate. These transient voltages must not exceed the rating of the device.

RMS Forward Current (I_F)—The I_F is the maximum steady-state current that the SCR can safely carry. If the construction of the SCR is such that a heat sink can be used, it must be used if the SCR is to carry the full rated current.

Surge Current (I_{SURGE})—Surge current is the maximum amount of current that the SCR can carry for a brief period of time. A brief period of time is usually understood to be the duration of a half cycle at 60 Hz, or about 8 milliseconds.

Forward on Voltage (V_F)—The V_F is the voltage drop that can be expected across an SCR when it is in the on condition and is operating within its specifications. When an SCR is used to switch current to a relay or an alarm, the power supply voltage must be great enough to handle the voltage drop across the SCR in addition to the voltage required by the load (relay or alarm).

Holding Current (I_H)—The I_H is the minimum current that must pass through the SCR to keep it turned on. The SCR should not be used with anode current below the rated holding current. In some cases, this might make it possible to turn off the SCR by removing the gate voltage, but the effect is not predictable, and operation of the circuit will be erratic.

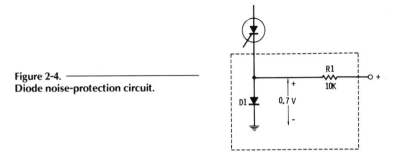

Figure 2-4.
Diode noise-protection circuit.

Other specifications are sometimes given in the manufacturer's literature, but those previously listed are usually all that need to be considered in designing most circuits.

Reducing Susceptibility to Noise

As pointed out earlier, an SCR will turn on if the gate is made positive for only about 5 μs. This makes it easy to trigger, but it also makes it vulnerable to being triggered by noise impulses. Usually, by careful design and construction, noise can be kept out of alarm circuits, but in some localities, the noise level is so high that some noise may get into the triggering circuits.

The boxed-in portion of Fig. 2-4 shows a scheme that will reduce the sensitivity of an SCR, and thereby reduce its vulnerability to noise. Diode D1 is a silicon diode of the type used as a power-supply rectifier. Resistor R1 assures that a small current will pass through the diode. It is a characteristic of a silicon diode that when there is current through it, there will be a small voltage drop—in the order of 0.7 V—across it. In Fig. 2-4, this voltage drop biases the cathode of the SCR positive by about 0.7 V. This means that any signal or noise impulse appearing between the gate and ground must be higher than this bias voltage before the SCR will turn on.

In many instances, the diode and resistor will not be necessary. In some localities, however, it may be necessary to use two diodes in series.

Inductive Loads

One of the SCR specifications listed above was the peak reverse voltage that could safely be applied. When the load is purely resistive, there will be no reverse voltage, but if the load is inductive, a spike of reverse voltage will be applied to the SCR whenever the circuit is interrupted.

Fig. 2-5 shows an SCR with an inductive load, such as a relay or a bell. Across the load there is a diode connected in the reverse direction. When a steady current is present in the load, there will be no current in the diode, because it is connected backwards. When the circuit is opened, however, the collapsing magnetic field in the inductive load will produce a voltage of reverse polarity across it. In Fig. 2-5, the reverse current passes harmlessly through the diode and does not damage the SCR.

THE UNIJUNCTION TRANSISTOR

Another component that is useful for intrusion alarms is the unijunction transistor (UJT) shown in Fig. 2-6. This device is a single bar of n-type silicon with two connections called base 1 and base 2. About halfway along the bar is another electrode, called the emitter, connected to the bar through a semiconductor junction. The silicon bar serves as a voltage divider between base 1 and base 2, making the voltage at the emitter junction about 60% of the voltage between the two bases. The emitter junction is back biased so that there is only a very small leakage current in the emitter circuit.

When the emitter voltage becomes more positive than the internal back bias voltage, emitter current begins. This current starts out high and decreases as the emitter voltage increases.

The UJT can be used as a time-delay generator to trigger an SCR. Fig. 2-7 shows a UJT pulse generator. When power is applied to the circuit, the voltage across capacitor C1 is zero, so the emitter current is practically zero. This capacitor charges through resistors R1 and R2. When the voltage across the capacitor becomes high enough, the UJT will "fire" and discharge the capacitor through resistor R4. This will produce a spike of voltage across R3, and the UJT will turn off. Capacitor C1 again charges, and the process repeats. The time duration between the pulses is proportional to the product of C1 and the sum of R1 and R2. The time period can be varied by varying R1. Resis-

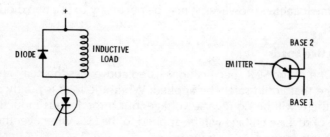

Figure 2-5.
Diode used with an inductive load.

Figure 2-6.
Unijunction transistor (UJT).

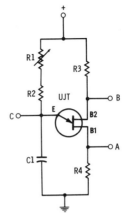

Figure 2-7.
Unijunction-transistor pulse generator.

tor R2 is provided to limit the emitter current to a safe value when R1 is set to its minimum position.

Although the normal output of the circuit of Fig. 2-7 is a positive-going pulse at point A, there is also a negative-going pulse available at point B. This is because the small current which normally appears between base 1 and base 2 increases when the UJT fires. A sawtooth voltage is available at point C. Any of these outputs may be used.

Fig. 2-8 shows a UJT connected to fire an SCR a fixed time interval after a circuit is opened. Normally the voltage across the capacitor is held at zero because switch S1 is closed. This switch might be a door-operated switch in an alarm system. When the switch is opened, capacitor C1 starts to charge. After a time delay—depending on the size of the capacitor and the series resistance—the UJT will fire

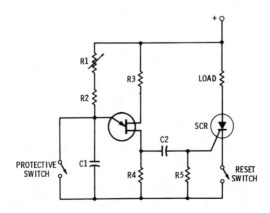

Figure 2-8.
UJT time-delay circuit.

and supply a positive-going pulse that can be used to trigger an SCR. This circuit has the obvious disadvantage that it will never fire the SCR if the door is opened and then closed quickly, but it serves to illustrate the operation of the UJT as a time-delay generator.

TYPE 555 INTEGRATED-CIRCUIT TIMER

One of the most versatile integrated circuits available is the Type 555, which was originally introduced by Signetics but is now available from many different integrated-circuit manufacturers. The designations used by various manufacturers for this IC are listed in Table 2-1.

Fig. 2-9 shows a schematic diagram of the integrated circuit. This diagram is not convenient for explaining how the circuit works, but it shows what is connected to each pin. By checking this diagram, you can get a good idea of whether or not some proposed circuit might result in damage to the IC.

Table 2-1. Designations of 555 Integrated Circuit

Manufacturer	Type Designation	
	General Purpose	Precision
Signetics	NE555	SE555
Intersil	NE555	SE555
Motorola	MC14555	MC1555
National	LM555C	LM555
Raytheon	RC555	RM555
Texas Instruments	SN72555	SN52555
RCA	CA555C	CS555

The operation of the circuit is better understood after observing the simplified block diagram in Fig. 2-10. Here we see that the device consists essentially of two comparators, a flip-flop, an output stage, and an output transistor.

There are really two outputs. One is the uncommitted collector of transistor Q1 at pin 7, and the other is the output at pin 3, which is the midpoint of a totem-pole arrangement. The condition of each output depends on the state of the flip-flop, which, in turn, is determined by the two comparators. The two states of the flip-flop are:

SET. When the flip-flop is set, the output at pin 3 is high, nearly at V_{cc}. The discharge transistor (Q1) is off, so its collector looks like an open circuit.

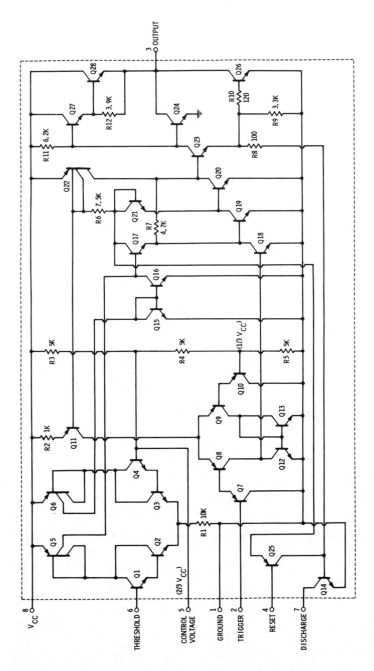

Figure 2-9.
Schematic diagram of the Type 555 timer. *(Courtesy National Semiconductor Corp).*

RESET. When the flip-flop is reset, the output at pin 3 is low, nearly at ground potential. Transistor Q1 is on, so pin 7 is, for all practical purposes, connected to ground.

The state of the flip-flop is determined by signals from the two comparators. As shown in the block diagram, one input of each of the comparators is connected to a voltage divider consisting of three 5K resistors in series. As we will see later, this arrangement makes the operation of the circuit nearly independent of the supply voltage. Before describing the operation of the circuit, we will discuss the function of each of the pins.

Pin 1. Ground.

Pin 2. Trigger. When the voltage at this pin falls lower than $^1/_3$ V_{cc}, the trigger comparator will provide a signal that will set the flip-flop.

Pin 3. This is the output pin. The load can be connected either between pin 3 and V_{cc} or between pin 3 and ground.

Pin 4. Reset. When this pin is not used, as in many circuits, it is connected to V_{cc}. If pin 4 is momentarily brought to ground, the flip-flop will be reset.

Pin 5. Control Voltage. A signal applied to pin 5 will change the normally fixed input voltages to the comparators. It can be used to modulate the duration or frequency of pulses gener-

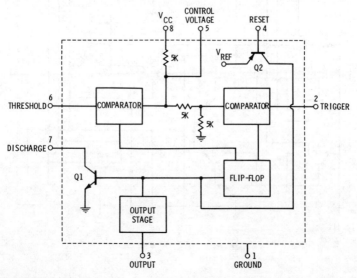

Figure 2-10.
Block diagram of the Type 555 timer.

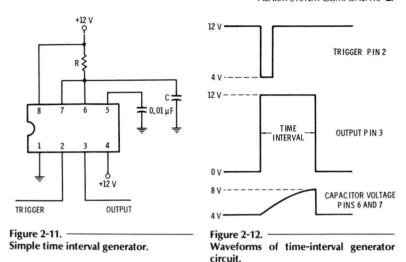

Figure 2-11. ————————————
Simple time interval generator.

Figure 2-12. ————————————
Waveforms of time-interval generator circuit.

ated by the circuit. When not used, this pin should be grounded through a 0.01 µF capacitor.

Pin 6. Threshold. When the voltage at this pin exceeds V_{cc}, the threshold comparator will provide a signal that will reset the flip-flop.

Pin 7. Discharge. This pin is connected to the uncommitted collector of transistor Q1.

Pin 8. V_{cc}. The supply voltage may range from $+4.5$ V to $+18$ V.

Many manufacturers offer a dual timer, often called Type 556, which consists of two Type 555 timers in a single integrated circuit.

In the following paragraphs, we will discuss the operation of the Type 555 timer by describing a circuit that we will use later in an actual intrusion alarm.

A SIMPLE TIMER

In Fig. 2-11 the Type 555 timer is used to generate a known time interval. Assume that initially the flip-flop in the timer is reset. The voltage at pin 3 is low, and transistor Q1 is turned on so that capacitor C cannot charge. When a negative-going pulse is applied to pin 2, the flip-flop will set. The voltage at pin 3 will go high, and transistor Q1 will turn off, allowing capacitor C to charge through resistor R. The capacitor will continue to charge until the voltage across it reaches $^{2}/_{3}$ V_{cc}. Inasmuch as the capacitor is connected to pin 6 (the threshold pin), when this voltage is reached, the flip-flop will reset.

The voltage at pin 3 will go low, and transistor Q1 will turn on, thus discharging the capacitor. The flip-flop will remain reset until the circuit is triggered again.

The waveforms associated with this operation are shown in Fig. 2-12. The time interval is started when pin 2 is brought negative. It ends when the output voltage at pin 3 goes to zero. The time interval is determined by the equation

$$t = 1.10RC$$

where,
    t is the time in seconds,
    R is the resistance in ohms,
    C is the capacitance in farads.

A plot of the time intervals that can be obtained with different values of resistance and capacitance is shown in Fig. 2-13. Resistor R should not be larger than 10 megohms. The value of the capacitor is limited by its leakage. Particular care should be exercised when using electrolytics to be sure that the leakage is low enough at the highest temperature at which the circuit will be used.

Inasmuch as the triggering levels of the comparators in the timer are fractions of the supply voltage rather than absolute values of voltage, the timer intervals will be nearly independent of variations in the supply voltage.

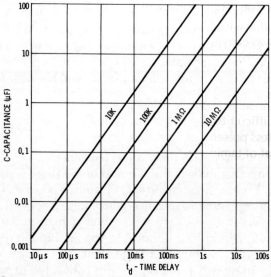

Figure 2-13.
Delay time for various values of R and C.

TAMING THE 555 TIMER

Any experimenter who has worked with the 555 timer has learned that although the time intervals are nearly independent of outside influences, the triggering of the timer may be very erratic indeed. The 555 timer is basically a very fast integrated circuit. A current of half a milliampere or less applied to pin 2 for a very short period of time will cause the circuit to trigger. This means that noise impulses of extremely short duration can cause erratic operation. Most circuits with which the experimenter is familiar are much slower in response. Normally, noise impulses of very short duration can be ignored. In fact, the average experimenter may not be aware that such noise impulses exist and are found in almost all equipment. By the same token, the output waveform of the 555 timer has very sharply rising and falling edges which can induce pulses in adjacent circuits.

Every wire or lead has some inductance. It may be only a fraction of a microhenry, but the inductance is present nevertheless. The voltage drop across an inductance is equal to the inductance in henries multiplied by the rate of change of current. Thus the inductance of a lead may be very small and the current may also be very small, but if the *rate of change of current* is high, the induced voltage will be high. We can get some idea of the magnitude of this effect by considering an actual situation.

Suppose that a lead in a circuit happened to have an inductance of 1 μH and was carrying a current of 100 mA. If the current dropped to zero in a tenth of a microsecond, a realistic time for a circuit as fast as the 555, the induced voltage would be 1 volt. This means that for a very brief instant, the voltage on the lead would be carrying a spike of 1 volt. The insidious part of these induced voltages is that they are so fast, it is difficult to spot them on an ordinary oscilloscope. They are nevertheless present and must be taken into consideration.

The secret of taming fast circuits like the 555 is to use bypassing judiciously. A ceramic-disc capacitor having a value of at least 0.05 μF should be connected between the power supply pins—pins 1 and 8—as close to the IC as practical. In addition, a good grade electrolytic capacitor of at least 10 μF should also be connected across these same two points. This capacitor need not be located right at the pins of the integrated circuit, but it should not be more than a few inches away.

The ceramic capacitor, which has very low internal inductance, will absorb sharp wavefronts, while the electrolytic will remove the

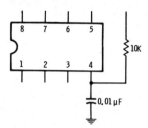

Figure 2-14.
Automatic reset circuit.

slower disturbances. Adding these two capacitors usually will tame the circuit so that it is rarely if ever influenced by noise on the power lines.

Another limitation of the 555 timer is that when it is connected as shown in Fig. 2-11, there is no way of being sure that the flip-flop will actually be in the reset state when power is applied to the circuit. This means that the circuit of Fig. 2-11 could be triggered when power is restored after a power failure.

This can be prevented with the arrangement of Fig. 2-14. Here a capacitor is connected between the reset pin—pin 4—and ground. When power is first applied to the circuit, the capacitor will be discharged, and pin 4 will be at ground potential. This assures that the circuit will be in the reset state. Shortly after power is applied, the capacitor charges to the supply voltage value, thus allowing the circuit to operate normally.

A SIMPLE PULSE TRANSFORMER

One of the problems that frequently occurs in building security systems is that a trigger pulse happens to have the wrong polarity to trigger another stage. Of course, this could be overcome by adding another transistor stage to invert the phase of the trigger pulse, but this complicates the system and adds to the power consumption.

Another solution to this problem is to use a pulse transformer that will invert a signal. Pulse transformers are commercially available, but often there may be no supplier in the experimenter's own community. An easy solution to the problem is to build the simple pulse transformer shown in Fig. 2-15. The core of this transformer is taken from a regular ferrite-rod antenna of the type used in small transistor radios. The characteristics of the core are not critical—any small ferrite core will usually work. The primary and secondary windings are identical. For the primary, about 65 to 100 turns of small—say number 26 to 30—wire are wound closely spaced on the core and held in place with plastic electrical tape. A layer of tape is placed

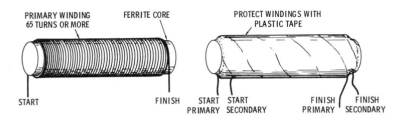

PRIMARY WINDING 65 TURNS OR MORE | FERRITE CORE | PROTECT WINDINGS WITH PLASTIC TAPE

START | FINISH | START PRIMARY | START SECONDARY | FINISH PRIMARY | FINISH SECONDARY

Figure 2-15.
Simple homemade pulse transformer.

over the winding. Then the secondary is wound over the tape in the same manner as the primary. Both windings are wound in the same direction. When the start of the primary winding goes high, the start of the secondary will also go high. The start and finish of each winding should be identified to simplify connecting the transformer in the circuit.

The little homemade transformer is not rated to carry any appreciable direct current. As a protective measure, the transformer may be capacitively connected into the circuit (Fig. 2-16A). In most cases, it will be satisfactory to connect a resistor in series (Fig. 2-16B) to limit the direct current.

The pulse transformer has an additional advantage in that it will provide a great deal of isolation between stages. This will make it easier to avoid ground loops and will usually improve noise immunity.

TYPE LM3909 LED FLASHER/OSCILLATOR

The Type LM3909 integrated circuit is a monolithic oscillator that was specifically designed to flash light-emitting diodes. The principal feature of this device is that it is designed to operate on a 1.5-V battery. By using an oscillator capacitor as a voltage booster, the device, powered by only a 1.5-V battery, can flash LEDs that won't oper-

0.1 μF | 1K.

(A) Capacitive coupling. | *(B) Series resistor.*

Figure 2-16.
Connections for simple pulse transformer.

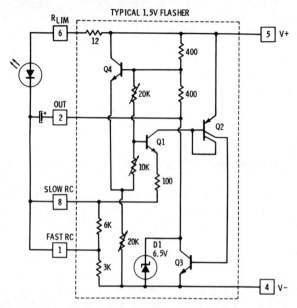

Figure 2-17.
Type LM3909 oscillator/flasher. (Courtesy National Semiconductor Corp.)

ate from a 1.5-V battery. If a low enough duty cycle is used, the battery will last for almost its shelf life.

Some of the features of the LM3909 make it ideal for use in security systems. It can be used with a battery power source to build infrared sources, eliminating the need to run any wires to the light source. It can also be used to generate alarm signals, and as an amplifier. The internal diagram of the device is shown in Fig. 2-17, and two flasher circuits are shown in Fig. 2-18.

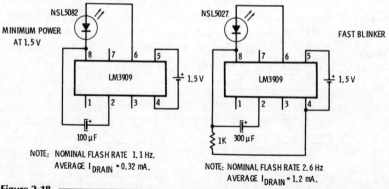

NOTE: NOMINAL FLASH RATE 1.1 Hz.
AVERAGE I_{DRAIN} = 0.32 mA.

NOTE: NOMINAL FLASH RATE 2.6 Hz
AVERAGE I_{DRAIN} = 1.2 mA.

Figure 2-18.
Typical LED flasher circuits using the type LM3909 flasher/oscillator.

ALARM TRIGGER CIRCUITS

CHAPTER 3 ——————————————————————————————

The heart of any intrusion alarm is a trigger circuit that will apply power to an alarm device, such as a bell, whenever there is an intrusion. The trigger circuits described in this chapter can be used in any type of intrusion-alarm system. They operate on a signal that is taken from some sort of protective circuit, and they apply power to the actual alarm device.

An intrusion alarm, by its very nature, must operate properly when there is no one around to control its operation. For this reason, the trigger circuit must be tailored to a particular operation. One feature that can be added to adapt an alarm to a particular application is the incorporation of time-delay circuits. For example, if a homeowner wants an alarm merely to wake him in the event of an intrusion, he can get by with a simple trigger circuit that will initiate an alarm. Of course, there should be a provision that will prevent the alarm from being turned off easily. For example, if an alarm is triggered by someone opening a window, it should not be possible to turn off the alarm by simply closing a window.

If, on the other hand, a homeowner uses an alarm to protect his home while he is away, he may well want an automatic shut-off circuit that will disable the alarm in the event of a false alarm or after a genuine alarm has been sounding for some time.

Inasmuch as time delay circuits are used with all sorts of alarms, they are covered separately in Chapter 10.

BASIC TRIGGER CIRCUIT

Fig. 3-1 shows a schematic diagram of a basic trigger circuit that can be used with any alarm system. In this circuit, the SCR will turn

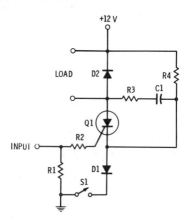

Figure 3-1.
Basic trigger circuit.

on whenever the input terminal is positive with respect to the cathode. Resistor R1 assures that the SCR will not turn on if the input circuit has some noise on it. Resistor R2 limits the current in the gate of the SCR to a safe value.

Diode D1 and resistor R4 are provided to increase the noise immunity of the circuit. These components are not needed in all alarms, but can be added if noise triggering is a problem. Resistor R3 and capacitor C1 form the snubber circuit that prevents the SCR from turning on when power is applied to the circuit. A snubber circuit should be included in any alarm system.

Diode D2, connected across the load, is necessary to protect the SCR against reverse-current spikes if an inductive load is used. It is not needed with a purely resistive load, but it would not do any harm.

The circuit in Fig. 3-1 turns on whenever the input goes positive, and it stays on until it is reset. The circuit can be reset by switch S1, which is a normally closed momentary-action switch—usually a push button.

This circuit is fine for many applications. Instead of a push-button switch for S1, a key switch or other activating arrangement can be used at this point to turn the circuit on and off. There are very few components in the circuit, so it can be made very reliable and should operate for many years with little trouble.

A parts list for Fig. 3-1 is given in Table 3-1. The resistors and capacitors in the circuit are not critical. Resistors with a ½-watt rating can be used. Capacitor C1 should preferably be a disc ceramic type,

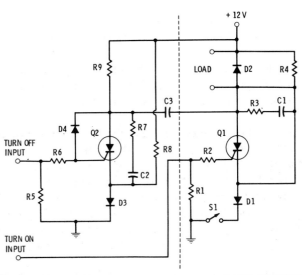

Figure 3-2.
Improved trigger circuit.

but any capacitor will be better than none. The diodes are silicon diodes such as those used for rectifiers in power supplies. They are not critical. A 1N5391 or almost any other silicon diode will work fine.

Table 3-1. Parts List for Fig. 3-1

Item	Description
C1	Capacitor, 0.05 μF, ceramic disc
D1, D2	Diode, silicon, 1N5391 (see text)
Q1	SCR, 2N3228 (see text)
R1	Resistor, 1K, ½ watt, carbon film
R2	Resistor, 330 ohms, ½ watt, carbon film
R3	Resistor, 10 ohms, ½ watt, carbon film
R4	Resistor, 10K, ½ watt, carbon film

Probably the most difficult component to obtain is the SCR. The parts list calls for a 2N3228, but almost any SCR that can carry the load current will do nicely. Probably the main precaution that must be observed is that diode D1, if used, must be able to carry the full-load current.

AN IMPROVED TRIGGER CIRCUIT

The trigger circuit of Fig. 3-1 is simple and dependable and it is adequate for many systems. The major limitation of the circuit is that once it is turned on, it can only be reset by manually opening switch S1. This means that the circuit is not suitable where it is desired to reset the circuit automatically with a timer. This limitation is overcome with the circuit of Fig. 3-2. Here one input can be used to turn the circuit on supplying power to the load, and another input can be used to turn the circuit off. A parts list for the circuit in Fig. 3-2 is given in Table 3-2.

That part of the circuit of Fig. 3-2 to the right of the dashed line is identical to the circuit of Fig. 3-1, and the components are designated accordingly. That portion to the left of the dashed line is used to turn SCR Q1 off in response to a positive-going pulse.

When power is first applied to the circuit neither of the SCRs will turn on. The circuit is thus armed and waiting for a positive pulse from the sensing circuits. Note that under this condition, both sides of capacitor C3 are at the same potential and the capacitor is not charged.

In a complete alarm system, the circuit is arranged so that in the event of an intrusion, the turn-on input will go positive. This will turn on SCR Q1 supplying power to the load—some kind of alarm device. When SCR Q1 turns on, the right end of capacitor C3 will drop to nearly ground potential. The left end is still at the supply voltage. This will cause capacitor C3 to become charged, with the left side positive and the right side negative. During this state, SCR Q1 will continue to conduct supplying power to the alarm device. It isn't necessary to keep the gate of the SCR positive.

Now, suppose it is desired to automatically turn SCR Q1 off and reset the circuit. For example, suppose that the alarm has been sounding for several minutes and that there has been no response. There is probably nothing to be gained by allowing it to continue to sound and if a battery supply is used the batteries will become discharged. A time switch can be provided to supply a positive pulse to the turn off input after some time interval.

When a positive pulse appears at the turn-off input the other SCR, Q2, will turn on. This will bring the left side of capacitor C3 to nearly ground potential. Now we have a charged capacitor with both ends connected together through ground. It will immediately discharge bringing the anode of SCR Q1 to ground potential. This will interrupt the current of SCR Q1 causing it to turn off.

Now we have the situation where SCR Q2 is on and SCR Q1 is off. The circuit is reset and is ready for another signal from the sensing circuits of the alarm system. Because of the high value of resistor R9, the circuit will draw very little current in this state. SCR Q1 can always be a small unit, even when a larger SCR is used to supply a heavier current to the alarm device.

Normally, with the component values specified in Fig. 3-2, capacitor C3 can be a 1.0-μF unit. However, if a heavier current unit is used for SCR Q1, it may be necessary to use a larger value for C3. Inasmuch as the polarity of the charge in C3 reverses during circuit operation, a regular electrolytic capacitor cannot be used. A nonpolarized 10-μF capacitor will handle any SCR that is practical in an intrusion alarm.

CONNECTING THE LOAD

The circuits of Figs. 3-1 and 3-2 will handle alarm devices that do not draw more than one ampere of current. There are two different ways to handle alarms that require more current. One way is to use a heavier SCR for Q1. If this is done, diode D1 and resistor R4 will probably not be required because most large SCRs are not very susceptible to noise.

Probably an easier way to handle large loads is to use a relay, with the coil connected to the load terminals in Fig. 3-1 or 3-2, and the contacts connected to the alarm device.

Table 3-2. Parts List for Fig. 3-2

Item	Description
C1, C2	Capacitor, 0.05 μF ceramic disc
C3	Capacitor, 1.0 μF (see text)
D1, D2, D3, D4	Diode, silicon 1N5391 (see text)
Q1, Q2	SCR, 2N3228 (see text)
R1, R5	Resistor, 1K, $\frac{1}{2}$ watt, carbon film
R2, R6	Resistor, 330Ω, $\frac{1}{2}$ watt, carbon film
R3, R7	Resistor, 10Ω, $\frac{1}{2}$ watt, carbon film
R4, R8, R9	Resistor, 10K, $\frac{1}{2}$ watt, carbon film
S1	Switch, N/C momentary action push button

A VOLTAGE-LEVEL TRIGGERING CIRCUIT

Fig. 3-3 shows an unusual triggering circuit that is triggered by changes in its own supply voltage. Usually, a triggering circuit is made immune to changes in supply voltage so that it will not be susceptible to false alarms. There are situations, however, where it is desirable to sound an alarm whenever the supply voltage changes. A good example of this is an automobile alarm. When the owner is

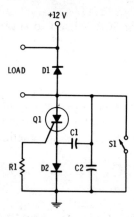

Figure 3-3.
Voltage-sensitive trigger circuit.

away from the automobile, there is no valid reason why there should be a sudden drop in the supply voltage from the battery. Anything that would cause such a change would be the result of someone tampering with the car. For example, if someone opens the door, turning on the dome lights, the voltage on the supply line will drop briefly. This will also occur if anyone tries to start the car.

In the circuit shown in Fig. 3-3, the SCR is turned on by the charge in capacitor C1. As pointed out previously, the voltage across a capacitor cannot change instantaneously. Thus, when the supply voltage drops, current will pass between the gate and cathode of the SCR, turning it on. The circuit can be reset by interrupting the current in the SCR. This can be accomplished by momentarily closing the normally open switch, S1.

This circuit is well suited to the protection of electrically operated devices that are otherwise hard to protect. The alarm system can be arranged to sound an alarm whenever a vehicle, such as a tractor or fork-lift truck, is started.

The parts list is given in Table 3-3.

The three simple trigger circuits described in this chapter will serve in any type of alarm. Their principal function is to apply power

Table 3-3. Parts List for Fig. 3-3

Item	Description
C1	Capacitor, 0.1 μF, 150 volts, paper
C2	Capacitor, 0.01 μF, ceramic disc
D1, D2	Diode, silicon, 1N5391
Q1	SCR, 2N3228
R1	Resistor, 100 ohms, ½ watt
S1	Switch, normally open, momentary action

to the alarm circuit. With these circuits the alarm will latch in the event of an intrusion and will stay energized until some shut-off action is taken, either manually with a switch or automatically with a timing circuit.

Of course, many other types of trigger circuits are available, but these circuits have the advantage of simplicity and reliability. They will work well under all reasonable conditions.

ELECTROMECHANICAL ALARMS

CHAPTER **4** ——————————————————————————

The electromechanical alarm is the simplest of all the alarm circuits available. People sometimes question its effectiveness. The fact is, however, that electromechanical systems can be made as effective as any other type of system, and they are simpler and hence more dependable.

One place where the ingenuity of the alarm constructor can really add to its effectiveness is in the selection or design of sensor switches. There are many different types of intrusion switches available commercially, and many more can be conceived by the technician.

A HOME-BUILT DOOR SWITCH

Fig. 4-1 shows a simple door switch that can be built in a few minutes. The switch consists of a small length of phosphor-bronze wire that slides between two eyelets connected to the trigger circuit. When the door is opened, the wire is pulled through the eyelets, and the circuit between the eyelets is opened (Fig. 4-1B), triggering the alarm. The phosphor-bronze wire is sometimes used for hanging pictures and can be obtained from a hardware store. The nylon cord can be ordinary fish line.

In spite of its simplicity, the arrangement can be very effective, particularly if it is concealed behind a drape, as shown in Fig. 4-2, where it will never be seen by the intruder.

WINDOW FOIL

One of the most common sensors used with electromechanical systems is a foil tape that is placed on store or home windows in such a way that the circuit will be opened if the window is broken. It is available either with or without an adhesive backing.

The first consideration in the use of window foil is to lay it out in a configuration that cannot easily be defeated by jumpering. For ex-

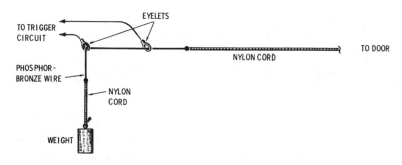

(A) Door closed.

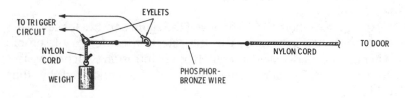

(B) Door open.

Figure 4-1.
Home-built door switch.

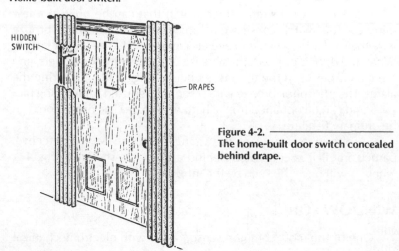

Figure 4-2.
The home-built door switch concealed behind drape.

ample, the arrangement of Fig. 4-3 is unsatisfactory because the foil obviously makes a complete circuit and the whole system can be defeated by simply connecting a jumper between points "A" and "B." A better arrangement is shown in Fig. 4-4. Here it is not apparent to the intruder how the circuit could be jumpered.

DOOR AND WINDOW SWITCHES

Many door switches, such as the plunger and magnetic types, are available. The plunger-operated switch shown in Fig. 4-5 is very popular. It is usually mounted flush in the hinged side of the door, where it will be concealed as shown in Fig. 4-6.

The magnetic switch shown in Fig. 4-7 is also very popular. The switch is mounted in the door jamb, and a small magnet is mounted in the door, as shown in Fig. 4-8. When the door is closed, the magnet holds the switch in the closed position. When the door is opened, the switch opens, setting off the alarm.

Another very popular magnetic switch is shown in Fig. 4-9. This switch consists of two parts that are enclosed in molded plastic. The part containing the magnet is attached to a door or window. The part containing the switch is mounted on the door jamb or window frame. Although it is easier to detect than the hidden magnet in Fig. 4-8, this switch is effective if it is located where an intruder cannot get at it.

PRESSURE-OPERATED SWITCHES

One of the most versatile switch arrangements for intrusion alarms is the pressure-operated switch. This device consists of a sealed rubber or plastic pad with a switch arranged so that it will be actuated whenever pressure is applied to the pad. This type of switch, as well as many other intrusion switches, is available either as normally open or normally closed.

Fig. 4-10 shows a switch mat and runner that may be concealed under a rug. It will be operated whenever anyone steps on the rug. The mat can be located anywhere protection is desired. For example, it might be located in a hallway, just inside or outside a door, or in front of a cabinet containing valuables. These mats are available in long runners for use in halls or in sizes suitable for use on stairs. A narrow version (Fig. 4-11) may be placed on a window ledge so that it will be triggered if an intruder attempts to enter through an open window.

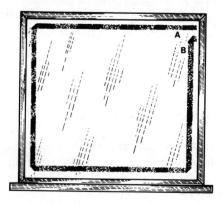

Figure 4-3.
Poor arrangement of window foil.

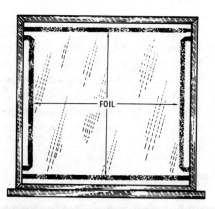

Figure 4-4.
Foil arranged so that circuit is not apparent to intruder.

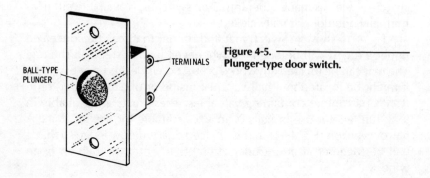

Figure 4-5.
Plunger-type door switch.

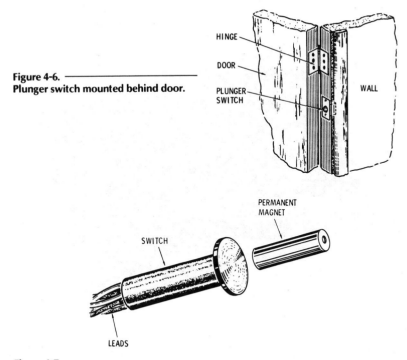

Figure 4-6.
Plunger switch mounted behind door.

HINGE

DOOR

PLUNGER
SWITCH

WALL

PERMANENT
MAGNET

SWITCH

LEADS

Figure 4-7.
Magnetic switch.

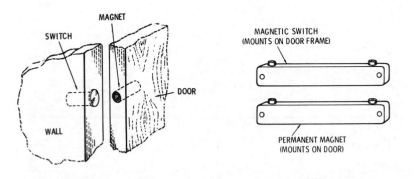

MAGNET

SWITCH

DOOR

WALL

MAGNETIC SWITCH
(MOUNTS ON DOOR FRAME)

PERMANENT MAGNET
(MOUNTS ON DOOR)

Figure 4-8.
Magnetic switch installed in door.

Figure 4-9.
Easy-to-mount magnetic door switch.

(A) Mat.

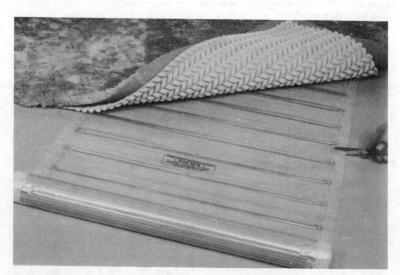

(B) Runner.

Figure 4-10.
Pressure switches. *(Courtesy Tapeswitch Corp. of America)*

A PET-PROOF PRESSURE SWITCH

A disadvantage of this type of pressure mat as an intrusion detector is that it can easily be triggered by a pet dog or cat roaming through the area. Fig. 4-12 shows a switching runner that is not subject to such false alarms. The sensor consists of several strips of pressure mats normally hidden under a rug. The strips are available in either the normally open or normally closed configuration; they are wired as shown in Fig. 4-13 so that two successive strips must be depressed at the same time to trigger the alarm. Because of the spacing of the strips, an animal will step on only one strip at a time, while a human being would step on at least two. Thus, the human intruder would trigger the alarm while the family pet would not.

Still another version of the pressure switch is shown in Fig. 4-14. This arrangement connects to an ordinary hose, so that the switch will be triggered whenever an automobile or other vehicle passes over the hose. Devices of this type are frequently used to detect automobiles in gas stations.

Other versions of the pressure switch are shown in Fig. 4-15. Some will trigger on one ounce of force or less.

FENCE PROTECTION SWITCHES

Fig. 4-16 shows a tension switch designed to trigger an alarm whenever a wire is stretched tightly. It can be used as a trip wire around an area or mounted on a wire fence so that it will trigger whenever anyone attempts to climb the fence.

The construction of the switch is shown in Fig. 4-17. A spring is arranged as shown to keep small pulls of the wire from throwing the switch. A strong pull will, however, stretch the spring and throw the switch.

This type of intrusion detector can be made portable and is very useful in protecting areas like temporary camp sites.

MISCELLANEOUS SWITCHES

The number of possible intrusion switches that can be built is limited only by the imagination of the constructor. Using a simple commercially available snap-action leaf switch, such as that shown in Fig. 4-18, switches can be built that will throw when an object is lifted or tilted or when a drawer is opened or closed.

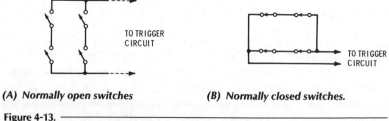

Figure 4-11.
Pressure switch for window ledge.

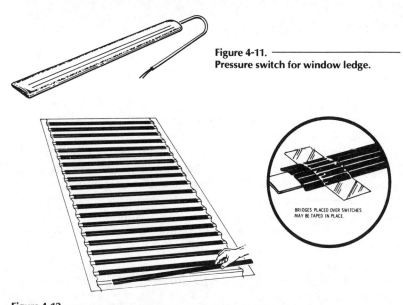

BRIDGES PLACED OVER SWITCHES
MAY BE TAPED IN PLACE.

Figure 4-12.
Pet-proof pressure switch. (*Courtesy Tapeswitch Corp. of America*)

TO TRIGGER
CIRCUIT

TO TRIGGER
CIRCUIT

(A) Normally open switches   *(B) Normally closed switches.*

Figure 4-13.
Wiring of pet-proof pressure switches (four strips shown).

Figure 4-14.
Pressure switch for use with hose.

(A) *Normally open.*

(B) *Momentary contact.*

Figure 4-15.
Miscellaneous pressure switches. *(Courtesy Tapeswitch Corp. of America)*

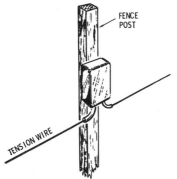

Figure 4-16.
Tension switch.

Figure 4-17.
Construction of tension switch.

Figure 4-18.
Snap-action leaf switch.

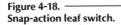

Figure 4-19.
Mercury switch.

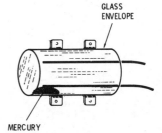

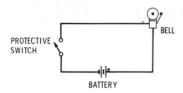

PROTECTIVE SWITCH

BELL

BATTERY

Figure 4-20.
A simple electromechanical alarm system.

Another basic switch that can be used as a sensor for an electromechanical intrusion alarm is the mercury switch shown in Fig. 4-19. This switch consists of two contacts that are closed through a small pool of mercury. This type of switch can be used to protect specific objects, or it can be arranged so that vibration will open the circuit and trigger the alarm.

ELECTROMECHANICAL INTRUSION ALARMS

The simplest electromechanical alarm consists simply of a switch, a bell, and a power supply connected in series as shown in Fig. 4-20. The switch could be mounted on a door so that the alarm would sound whenever the door was opened.

This circuit has several limitations. First, the alarm only sounds while the switch is closed. If, for example, the switch is mounted on a door, the bell will only sound while the door is open. A burglar could enter the premises quickly, closing the door behind him, and the alarm would only sound for a few seconds. Furthermore, an alarm of this type is not "fail-safe"; that is, if any of the components like the bell or the wiring should open or if the battery should fail, the alarm would not sound, and the owner would be unaware that he was unprotected. In addition, the alarm could be easily defeated by cutting the wires at any point.

Although the simple electromechanical alarm is not normally used as an intrusion alarm, it is widely used as an annunciator to give a warning whenever anyone enters an area such as the doorway of a store.

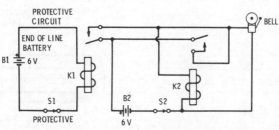

PROTECTIVE CIRCUIT

END OF LINE BATTERY

B1 6 V

K1

BELL

K2

S1

B2

S2

PROTECTIVE

6 V

Figure 4-21.
A relay-type intrusion alarm.

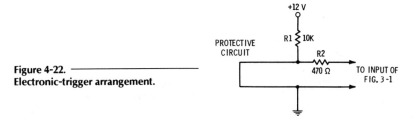

Figure 4-22.
Electronic-trigger arrangement.

A Relay-Type Intrusion Alarm

A much better intrusion alarm is shown in Fig. 4-21 (parts list in Table 4-1). In this circuit, battery B1 keeps a small current flowing in the protective circuit at all times. Under this condition, the contacts of relay K1 are held open. Therefore, no current flows in the coil of relay K2 or the alarm bell. When an intruder breaks the protective circuit in any way, including opening switch S1, relay K1 is de-energized and its contacts close. This completes the circuit to the bell and to relay K2. The function of relay K2 is to "lock" the circuit on so that the bell will continue to operate even if the protective circuit is closed. The bell will now continue to ring until the system is reset by momentarily opening switch S2. It cannot be reset by merely closing the protective circuit.

Table 4-1. Parts List for Fig. 4-21

Item	Description
B1, B2	Battery, 6 volts
Bell	Doorbell, 6 volts
K1	Relay, 6 volts, low current, normally closed
K2	Relay, 6 volts, normally open
S1	Switch, protective
S2	Switch, spst

This circuit is easy to construct and will provide good protection for a home or business. However, its limitation is that the relays, after prolonged use, require maintenance to prevent the ill effects of dust and corrosion. Furthermore, the battery drain is much higher than in the electronic systems described in the following paragraphs.

Electronic Trigger Circuit

The electromechanical intrusion detector is well suited for use with the electronic trigger circuits described in Chapter 3. Probably

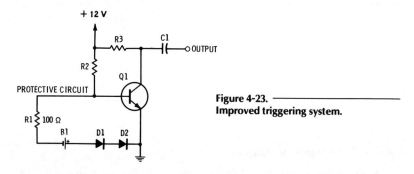

Figure 4-23.
Improved triggering system.

the simplest way to connect an electromechanical detector to a trigger circuit is shown in Fig. 4-22. Here the output is kept at ground potential because it is shorted through the protective circuit. When the protective circuit is opened, the output voltage rises. The output can be connected directly to the input of the trigger circuit shown in Fig. 3-1. With this arrangement, the alarm is triggered whenever the protective circuit is opened.

If the trigger arrangement is to be used with an alarm that has an automatic shutoff feature, it should be capacitively coupled. Otherwise, the alarm will continue to be retriggered as long as the protective circuit is open.

The arrangement in Fig. 4-22 has the advantage that if the wires leading to the protective circuit are cut, the alarm will be triggered. It does not respond, however, if the wires leading to the protective circuit are shorted together. The circuit to be described next provides this protection.

An Improved Electronic Trigger Circuit

Fig. 4-23 shows an improved electronic trigger circuit. Here a battery at the end of the protective circuit keeps transistor Q1 turned off. Its collector is thus at the positive power-supply voltage. If the protective circuit is opened, transistor Q1 is turned on by the base current through resistor R2. As the transistor turns on, its collector voltage will drop suddenly. This causes a negative-going trigger pulse at the output.

This circuit can be used directly with any circuit that requires a negative-going pulse to trigger it. We will describe some timing circuits of this type in Chapter 5. It can also be used with the trigger circuits of Figs. 3-1 or 3-2 if a pulse transformer of the type described in Chapter 2 is used to invert the polarity of the output pulse.

Table 4-2. Parts List for Fig. 4-23

Item	Description
B1	Battery, 3 volts, two Type A cells in series
C1	Capacitor, 10 μF, 35 volts, electrolytic
D1, D2	Diode, 1N5371, silicon
Q1	Transistor, any small-signal npn type
R1	Resistor, 100 ohm, $\frac{1}{2}$ watt
R2	Resistor, 47K, $\frac{1}{2}$ watt
R3	Resistor, 10K, $\frac{1}{2}$ watt

Note that this circuit cannot be foiled by shorting the protective circuit. Shorting this circuit amounts to connecting the base of the transistor to the anode of diode D1. Inasmuch as diodes D1 and D2 are silicon diodes, the voltage drop across them will be about twice the normal base-to-emitter voltage of the transistor. Thus current will be present in the base circuit of the transistor, turning it on.

The components in this circuit are not critical. The author has not found a small-signal silicon npn transistor that has not worked in the circuit. A list of parts is given in Table 4-2.

An alternative arrangement that provides a positive-going pulse in the event of an intrusion is shown in Fig. 4-24. Here a pnp transistor is used, and the positive side of the power supply is grounded. Normally the battery B1 keeps the transistor off so that its collector is at the potential of the negative supply voltage. If the protective circuit is opened, base current through resistor R2 pulls the base of the transistor negative, turning it on. The collector will drop to nearly ground potential, which is positive compared to the negative supply. This will cause a positive going pulse to appear at the output.

Note that with one exception, the parts are identical to those listed in Table 4-2. The exception is the transistor which is now a pnp

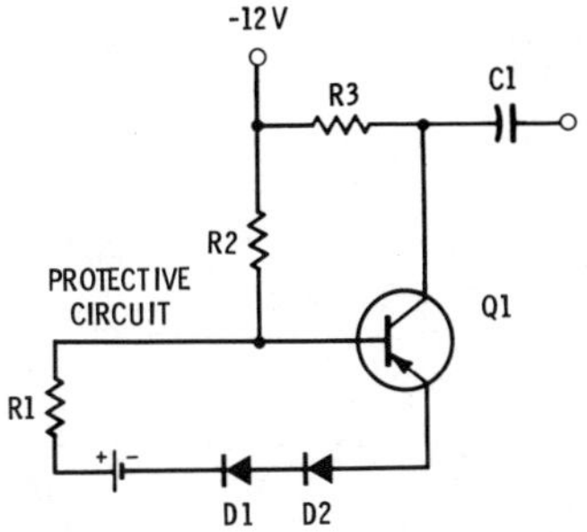

Figure 4-24.
Triggering system providing a positive-going pulse.

type. Again, almost any small- signal transistor will operate in the circuit. A type 2N3906 has been tried and found satisfactory.

In this day of many sophisticated advances in the field of electronics, the electromechanical intrusion alarm is often thought of as old fashioned. It is certainly old, being the first type of electrical intrusion alarm, but it is certainly not outmoded. The simple electromechanical alarm is still the most reliable of all types of alarm systems. It is rarely affected by any outside influences and has a very high reliability.

The effectiveness of this type of alarm depends on the ingenuity of the person installing it. When properly installed, it is nearly impossible to defeat.

PHOTOELECTRIC INTRUSION ALARMS

CHAPTER 5 ——————————————————————

The photoelectric intrusion alarm uses visible or infrared light to detect the presence or motion of an intruder. Its principal advantage is that a beam of light can be directed into areas where it would be impractical, if not impossible, to install wiring. Inasmuch as a photoelectric alarm is more complex than a simple electromechanical alarm, it requires more care in construction and installation to assure a reliable system.

The earliest photoelectric alarms were configured as shown in Fig. 5-1. Here a beam of light is directed across a path that is to be protected, such as a doorway. At the other side of the path a photosensitive device detects the presence of the light beam. If an intruder crosses the path, the beam will be interrupted. The photosensitive device detects the break in the light beam and initiates an alarm.

The simple system of Fig. 5-1 has many limitations. Unless great care is taken, the beam of light may be visible to an intruder who can then step over or crawl under it. The alarm can be foiled completely by simply shining an ordinary flashlight into the photosensitive device. This will prevent the alarm from being tripped, regardless of

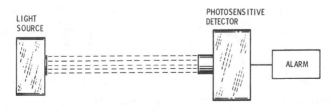

Figure 5-1. ——————————————————————
Arrangement of basic photoelectric alarm.

(A) Pictorial.

(B) Schematic symbol.

Figure 5-2.
Phototransistor (no base connection).

whether or not the light beam is interrupted so that an intruder can enter the area undetected.

Recent advances in solid-state technology have made infrared sources and receivers available at reasonable cost, so modern alarm systems rarely use visible light. This makes it much more difficult for a prospective intruder to spot the system. Modern systems often use modulated light beams that make it impossible to foil the system with a flashlight.

THE PHOTOTRANSISTOR

The photosensitive device used in the alarms described in this chapter is a phototransistor. This device, shown in Fig. 5-2, has two or three terminals and consists of a transistor with a built-in lens that will focus light on the base region. Normally only leakage current flows between the emitter and collector. This current will increase greatly when light falls on the base region. Some phototransistors have a base lead. Any connection to this lead reduces the sensitivity to light; therefore, it is not used in any of the circuits in this book.

Phototransistors are very small and may be used in applications where they can easily be concealed. It is possible to get phototransistors that respond to visible light or to infrared radiation. The latter is normally used where the beam is to be invisible.

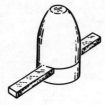

(A) Pictorial.

(B) Schematic symbol.

Figure 5-3.
Light emitting diode (LED).

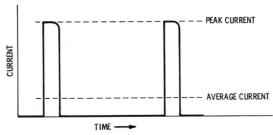

Figure 5-4.
Pulsed current for flashing an LED.

THE LIGHT-EMITTING DIODE (LED)

The LED, or solid-state lamp, as it is sometimes called, is a two-terminal device that consists of a pn junction which emits light when it is biased in the forward direction (Fig. 5-3). Units are available that emit in a wide range of wavelengths ranging from visible light to infrared radiation. They can be used in conjunction with phototransistors to provide very effective intrusion alarms.

AN LED DRIVER CIRCUIT

An LED will operate whenever a dc current passes through it. Thus it is possible to drive an LED with a simple dc supply. The reason that this is rarely done is that the average amount of light or infrared emitted by the LED will be quite small. This small light will require either a very sensitive receiver, or a relatively short path to be effective in an alarm system.

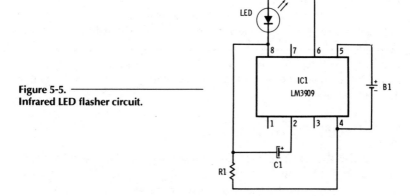

Figure 5-5.
Infrared LED flasher circuit.

A much better way to drive an LED in an intrusion alarm is to use a pulse source. The waveform of the current through the LED will be as shown in Fig. 5-4. Here it can be seen that the peak current can be quite high, while the average current is low. This means that the LED will emit strong pulses of light or infrared without excessive power dissipation. The Type LM3909 integrated circuit which was designed specifically as an LED flasher is ideal as a source.

Fig. 5-5 shows the circuit of a very good LED driver. (The parts list is given in Table 5-1.) It is very simple, requiring only a resistor and a capacitor in addition to the LED and IC. It also has the advantage that due to the low average power of the circuit, it will operate for months on a single 1.5-V type D dry cell. This is a blessing in many applications where wiring to the source would be difficult.

Table 5-1. Parts List for Fig. 5-5

Item	Description
B1	Dry cell, 1.5 volts, Type D
C1	Capacitor, 330 μF, 16 volts, electrolytic
R1	Resistor, 1K, ½watt, carbon film
IC1	LM3909 integrated circuit

The circuit of Fig. 5-5 will provide about three or four flashes per second. The circuit is not designed for high frequency stability and under extreme conditions of temperature and battery voltage the flash rate may change by as much as 50%. This is rarely a problem in well designed systems.

The principal difficulty with the circuit of Fig. 5-5, which uses an infrared LED, is determining whether or not it is working properly. The emission from the LED is not visible to the eye, so you can't tell whether or not it is working by looking at it. About the only thing that will keep the circuit from working is excessive leakage in the LED.

Probably the best way to check the circuit is to build the receiver first. The receiver is easier to check and can then be used to check the transmitter.

A PHOTOELECTRIC RECEIVER

Fig. 5-6 shows the schematic diagram of a photoelectric receiver that is designed to be used with the circuit of Fig. 5-5. The photosensitive device is a phototransistor with a peak sensitivity in the infra-

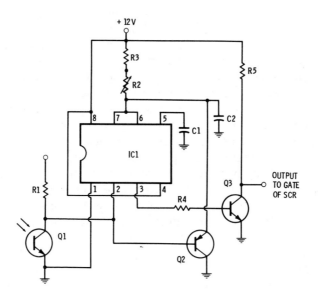

Figure 5-6.
Infrared receiver for use with flasher of Fig. 5-5.

red portion of the spectrum. The phototransistor can be thought of as a variable resistor. When it is dark, its resistance is very high and there is almost no current between the emitter and collector. Thus the voltage at the collector is very nearly equal to the supply voltage.

When a pulse of infrared from the LED reaches the phototransistor its resistance drops sharply. Current will now flow between the emitter and collector, and the collector voltage will drop. Thus, the output of the phototransistor is a series of negative-going pulses that correspond to bursts of infrared from the driver.

Table 5-2. Parts List for Fig. 5-6

Item	Description
C1	Capacitor, 0.01 Mylar
C2	Capacitor, 4.7 µF, 35 volts, electrolytic
1C1	Timer, Type 555
Q1	Phototransistor
Q2	Transistor, pnp type 2N3906
Q3	Transistor, npn type 2N2222
R1	Resistor, 47K, ½Watt, carbon film
R2	Potentiometer 100K, linear taper
R3, R5	Resistor, 22K, ½Watt, carbon film
R4	Resistor, 1K, ½Watt, carbon film

The rest of the circuit is what is commonly called a missing pulse detector using a Type 555 timer. The timer is set for a time interval somewhat longer than the time between the pulses from the phototransistor. The first pulse from the phototransistor will start the circuit timing and the output at pin 3 will go to the high state. If no further pulses were received, the circuit would soon complete its timing interval and pin 3 would go to the low state. However, before the time interval has elapsed, another pulse will arrive starting the timing cycle all over again. Thus the output at pin 3 will remain in the high state. If the stream of pulses is interrupted, the circuit will complete the timing cycle before being reset and the voltage at pin 3 will drop to the low state. The following stage, consisting of transistor Q1 and its associated components, merely inverts the signal at pin 3 of the timer. Thus the output of the complete circuit remains in a low state as long as the pulses from the phototransistor arrive regularly. If a pulse is missing, the output will go high. This positive-going pulse is fed to one of the trigger circuits of Chapter 3 to initiate an alarm.

The duration of the timing cycle is set by adjusting potentiometer R2. It should be set so that the receiver will respond fast enough for a particular application. If the cycle is too short, the circuit will trip if the pulse rate from the LED driver changes slightly. The cycle should be long enough so that it will tolerate a small change in the flash rate of the LED.

The circuit of Fig. 5-6 is rather easy to troubleshoot because most IR phototransistors are sensitive to visible light. By connecting a voltmeter between pin 3 of the timer and ground, and flashing a light at the phototransistor, you can verify that the timing cycle is started.

The housing for the receiver should include some means of shielding the phototransistor from ambient light. Usually a small metal tube, painted black on the inside, will provide adequate shielding.

The circuit of Fig. 5-6 is adequate for many systems. It has the disadvantage that the phototransistor is directly connected to the timer. This means that changes due to ambient light and temperature will also be coupled directly to the timer. It further means that the sensitivity of the circuit is limited.

A COMPLETE PHOTOELECTRIC SYSTEM

Fig. 5-7 illustrates the principle of using the LED driver of Fig. 5-5 and the receiver of Fig. 5-6. By far the most difficult part of installing and adjusting the system is properly aiming the light beam at the re-

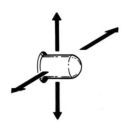
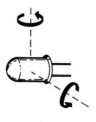

(A) Horizontal or vertical movement.   *(B) Movement along two axes.*

Figure 5-7.
LED and phototransistor can be moved in position or orientation.

ceiver. On paper the task looks simple, but this is deceiving. As shown in Fig. 5-7 both the LED and the phototransistor can be moved in position and orientation. That is, either one can be moved either horizontally or vertically, as in Fig. 5-7A, or along two axes as in Fig. 5-7B. Making the proper moves at either end of the path can be very frustrating. It sometimes helps to temporarily substitute a visible light LED for the infrared unit so that the direction of the beam can be determined visibly.

The path of the system should be selected so that an intruder cannot cross the beam in the interval between pulses. Usually this isn't a problem because the intruder has no way of knowing when a pulse is present. As shown in Fig. 5-8, an intruder remains in a diagonal beam longer than in a perpendicular beam. In an area such as a

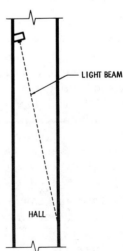

LIGHT BEAM

HALL

Figure 5-8.
Intruder will remain in a diagonal beam longer than in a perpendicular beam.

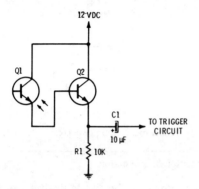

Figure 5-9.
A light detector circuit.

long hall where an intruder might move very rapidly it is sometimes advantageous to orient the system so that the beam is diagonal to his direction of travel.

Incidentally, this points out the importance of keeping the principle of the operation of a system secret. Whereas a pulsed infrared system might possibly be foiled by an intruder moving across the path very rapidly, another type of system, such as an audio alarm or vibration alarm, might be foiled if the intruder moves very slowly. If the intruder has no idea of the principle of the operation of the system, he will not know how to foil it.

A LIGHT DETECTOR

Fig. 5-9 shows the circuit of an alarm that will trigger whenever light falls on the phototransistor. It is useful in alerting someone whenever light is present in a protected area. For example, if an intruder were to turn on a flashlight in the protected area, the alarm would be triggered.

The parts list for Fig. 5-9 is given in Table 5-3.

An interesting application of this circuit is shown in Fig. 5-10. Here, another type of alarm is used to detect an intrusion at a remote point. In addition to setting off an alarm, the system also flashes a light such as a photoflash lamp. The light from the photoflash is picked up by the light detector which in turn triggers a local alarm.

This arrangement makes it possible to trigger an alarm at a location several feet away without using wires that might be cut by an intruder.

The circuit of Fig. 5-9 is used in this application because it is sensitive to light level changes rather than to absolute light level. The

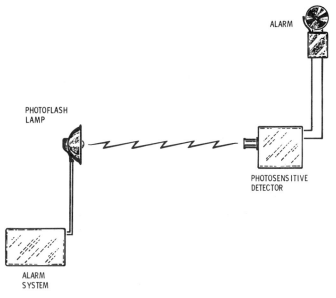

Figure 5-10.
Photoelectric remote alarm.

phototransistor must be sufficiently protected from ambient light so that it will not be saturated.

Table 5-3 Parts List for Fig. 5-9

Item	Description
C1	Capacitor, 10 μF, 35 volts, electrolytic
Q1	Phototransistor
Q2	Transistor, npn, Type 2N2222
R1	Resistor, 10K, ½ watt

The circuit of Fig. 5-9 can also be used with the receiver of Fig. 5-6. The output is then connected to pin 2 of the Type 555 timer and to the base of transistor Q2. With this arrangement, the circuit will respond only to changes in light level and not to ambient light.

A FOOLPROOF PHOTOELECTRIC DETECTOR

One of the major limitations of most photoelectric intrusion alarms is that they can be defeated by shining a light on the photodetector. That is, an intruder can focus a flashlight on the de-

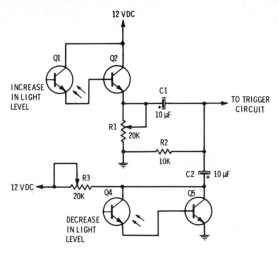

Figure 5-11.
Circuit that detects any change in light level.

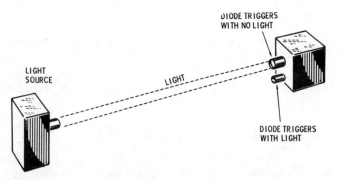

Figure 5-12.
Pictorial diagram of circuit that detects any change in light level.

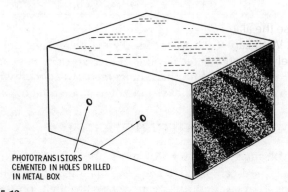

Figure 5-13.
Phototransistors installed in metal box.

tector before he breaks the light beam. The light from the flashlight will keep the alarm from triggering.

Figs. 5-11 and 5-12 show a system that cannot be defeated in this way. It actually consists of two systems. One triggers the alarm when the light level decreases, and the other triggers the alarm when the light level increases. Thus, any normal interruption of the beam will trigger the alarm, but it will also be triggered if anyone tries to defeat it by shining a light on the phototransistor. The parts list for Fig. 5-11 is given in Table 5-4. When installing this system the light source should be focused on the detector that triggers when the light level decreases. The tubes protruding from the box in Fig. 5-13 should be approximately ¼ by ¼ inch. After the system has been installed, R1 and R3 should be adjusted for proper operation. These potentiometers control the sensitivity of the system. The proper amount of light that will trigger the alarm can be preset by R1; then R3 can be adjusted so that the required decrease in light intensity will trigger the alarm.

Table 5-4. Parts List for Fig. 5-11

Item	Description
C1, C2	Capacitor, 10 μF, 35 volts, electrolytic
Q1, Q4	Phototransistor
Q2, Q5	Transistor, npn, Type 2N2222
R1, R3	Potentiometer, 10-20K
R2	Resistor, 10K, ½ watt

The circuit of Fig. 5-11 can also be used to monitor an area (like a room) for variations in light level. The phototransistors can be arranged as shown in Fig. 5-13 (the author used epoxy cement to hold them in place) so that they will view the area to be protected. In almost every case the alarm will be triggered if the lights are turned either on or off. With careful adjustment, the circuit will trigger when someone merely walks through the protected area.

PROXIMITY ALARMS

CHAPTER 6 ————————————————————————————

A proximity alarm is one that sounds whenever an intruder touches or comes close to the protective circuit. The protective circuit, usually a metal plate or a wire, acts as one plate of a capacitor with the ground acting as the other plate. An intruder approaching the wire will change the capacitance to ground. The trigger circuit detects this change in capacitance and triggers the alarm.

One convenient way of detecting a change in capacitance is to use the capacitor in the tuned circuit of an rf oscillator and detect changes in the oscillator frequency. Many existing proximity alarms operate on this principle. Unfortunately, this arrangement tends to radiate and cause rf interference. Current FCC regulations are severely restrictive of this type of alarm, and for this reason the proximity alarms described in this chapter do not use rf energy.

PRECAUTIONS

Before going into the details of proximity alarms, a word of warning is in order. Since a proximity alarm operates by detecting a change in its environment, it is very sensitive to outside influences like stray pickup from power lines and lightning discharges. This sensitivity increases as the alarm is made more sensitive. For this reason, a proximity alarm should not be any more sensitive than necessary to provide protection.

Probably no two proximity alarms will have protective circuits that are exactly the same. For this reason, it is difficult to design a proximity alarm that will work well in all possible installations. The alarms described in this chapter have all been tested, but it is likely

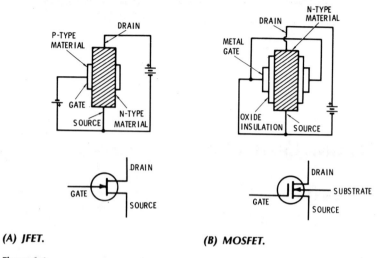

(A) JFET.   **(B) MOSFET.**

Figure 6-1.
Field-effect transistors.

that slight changes may be necessary to adapt them to a particular application.

THE FIELD-EFFECT TRANSISTOR

The field-effect transistor has an extremely high input impedance. Because of this, it can be used to detect small changes in capacitance and can be used in a proximity alarm. There are two basic types of field-effect transistors—the junction type and the insulated-gate type.

Fig. 6-1A shows the construction and schematic symbol of a junction FET (JFET) in simplified form. It consists basically of a "channel" of semiconductor material through which current can flow freely. The terminals at the ends of this channel are called the source and drain. They are analogous to the cathode and plate, respectively, of a triode vacuum tube.

The channel of conducting material is surrounded by a semiconductor material of the opposite type. This material, called the gate, forms a pn junction with the channel and is analogous to the grid of a triode tube. When a voltage of the proper polarity is applied to the gate, current will not enter the channel because the junction is reverse biased. The voltage will, however, constrict the channel and reduce the current between the source and drain. Thus, the FET acts very much like a vacuum tube triode, since a voltage applied to the gate controls the magnitude of the current between the source and drain.

The insulated-gate FET, shown in Fig. 6-1B, operates in a similar way. The difference is that instead of using a semiconductor diode between the gate and channel, the gate is actually insulated from the channel by a thin layer of oxide. For this reason, insulated-gate FETs are frequently called MOSFETs, where MOS stands for metal-oxide semiconductor. The MOSFET has an even higher input impedance than the junction type; therefore, it is the kind predominantly used in proximity alarms.

There is, however, one precaution that must be observed when using MOSFETs. Because of the extremely thin layer of oxide insulation that can easily be damaged by high voltages or static charges, they are shipped with the leads shorted together so they will not be damaged by static charges that might occur in transit. The leads should be kept shorted until they are connected into the circuit.

A MOSFET PROXIMITY ALARM

The principle of operation of a MOSFET proximity alarm is shown in Fig. 6-2. Here the protective circuit is a wire that acts as one plate of a capacitor with the ground or chassis acting as the other plate. The wire is connected to the gate of the MOSFET, but since the gate is insulated from the channel, any accumulated charge has no place to go and will remain on the wire.

The voltage across a capacitor is given by the following equation:

$$E = \frac{Q}{C}$$

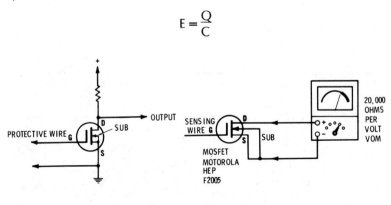

Figure 6-2.
Principle of MOSFET proximity alarm.

Figure 6-3.
Circuit for demonstrating proximity effect.

where,

> E is the voltage in volts,
> Q is the charge in coloumbs,
> C is the capacitance in farads.

If the charge remains relatively constant, anything that changes the capacitance of the capacitor will also change the voltage across its terminals. Thus, almost any object coming close to the protective wire will change its capacitance and hence the voltage at the gate. This in turn will change the current through the channel.

The voltage across the capacitor will also change if a charged object is brought close to it. Usually a person walking around will accumulate a charge, and this will add to the sensitivity of the alarm.

The principle of the MOSFET proximity alarm can be demonstrated with the arrangement of Fig. 6-3. Here an ohmmeter, set to the "ohms × 1K" range, is connected between the source and drain of a MOSFET. A sensing wire is connected to the gate. Whenever any object is brought close to the sensing wire, the indication of the ohmmeter will change significantly. Charged objects should not touch the sensing wire, since this might damage the MOSFET.

This arrangement may also be used to check the effectiveness of any proposed proximity alarm installation. The sensing wire can be installed in position and connected to the ohmmeter, as shown in Fig. 6-3. Then have a person walk through the protected area, and notice how much change there is in the ohmmeter reading. For an effective system, the change in the ohmmeter reading should be as great as possible.

The arrangement will also show whether a proposed installation will be subject to false alarms from outside influences. Simply connect the ohmmeter to the sensing wire and watch the indication. If

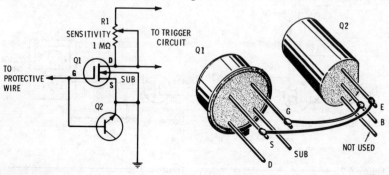

Figure 6-4.
Practical MOSFET proximity alarm.

Figure 6-5.
Method for protecting FET gate.

Figure 6-6.
Using ground wire with outdoor proximity system.

the pointer swings widely when nothing is moving in the area, the alarm is probably being influenced by stray pickup. In such cases, it might be advisable to use some other type of alarm.

A practical circuit for a proximity alarm is shown in Fig. 6-4 (parts list in Table 6-1). Transistor Q2 does not normally enter into the operation of the circuit. It merely serves to protect the FET from damage that it might encounter if the protective circuit were touched by someone carrying a charge. This transistor should be connected into the circuit across the FET, as shown in Fig. 6-5, as soon as the FET is installed. It will then protect the FET from damage.

The success of the system depends upon how carefully it is installed. The protective wire cannot run close to power lines and should be no longer than necessary to provide adequate protection. The insulation supporting the protective wire should be very good because any leakage will reduce the sensitivity. The input impedance of the FET may be over 100 megohms, and the circuit should be well insulated. This becomes more important as the wire is made longer.

This circuit works very well in indoor applications where the protective circuit is only a few feet long. But it is limited in outdoor applications where it can be triggered by ambient noise like thunderstorms. The arrangement of Fig. 6-6, where a ground wire is run above the sensing wire, will help minimize this effect.

Table 6-1. Parts List for Fig. 6-4

Item	Description
Q1	MOSFET, RCA Type 40559 or equivalent
Q2	Transistor, silicon, npn, 2N2713 or Motorola HEP54
R1	Potentiometer, 1 megohm, linear taper

It is very easy to be deceived into believing that this circuit is operating properly when it is actually operating on 60-Hz pickup from a

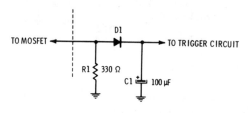

Figure 6-7.
Filtering circuit for MOSFET proximity alarm.

power line. An intruder approaching the protective circuit in a strong field will increase the amount of 60-Hz energy picked up by the system and will trigger the alarm. There is nothing wrong with this arrangement if the builder knows that it will not be as sensitive when the power fails. If foolproof operation is required, the system should be tested when the power in the building is turned off.

FILTERING AND AVERAGING

The circuit of Fig. 6-4 can be improved by adding the rectifier and filter of Fig. 6-7. This circuit rectifies the variations in the signal and averages them. This tends to minimize the influence of stray pickup, radio-station signals, and miscellaneous static discharges.

The amount of filtering depends on the values of capacitor C1 and resistor R2. The values listed in Table 6-2 may be adjusted experimentally for best performance in a particular installation.

Table 6-2. Parts List for Fig. 6-7

Item	Description
C1	Capacitor, 100 µF, 25 volts, electrolytic
D1	Diode, silicon, 1N4003
R1	Resistor, 330 ohms, ½ watt

A WORD OF CAUTION

The proximity alarm can cause more problems than any other alarm described in this book. This does not mean that good proximity alarms cannot be built. They can, but it is usually a cut and try process. However, the building and installing of such an alarm should be thought of as more of an experimental and design project than a simple construction project.

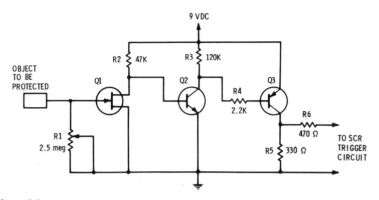

Figure 6-8.
A "touch alarm."

The biggest problem is that so many different factors can influence the voltage on a sensing wire. After all, the sensing wire would probably work quite well as a receiving antenna. The goal is to make the influence of an intruder different from other influences that might trip the alarm. Simply decreasing the sensitivity of the system may help to some extent, but the static charges associated with thunderstorms can induce very high voltages on the sensing wire.

Usually the most successful way of minimizing influences is to use the ground wire of Fig. 6-6. It is important that the end of the ground wire be connected to the circuit right at the ground point of the FET. The idea behind this arrangement is that most things will affect the grounding wire and the sensing wire in the same way, whereas an intruder will affect the capacitance *between* the wires. Thus a nearby lightning stroke should induce the same voltage in both of the wires and not produce a voltage between them.

Unfortunately, whether or not the circuit actually works this way depends heavily on the geometry of the installation and this is usually not the same in any two installations. Two wires that appear superficially to follow the same geometry may be just unsymmetrical enough to cause problems.

The best way to approach the proximity system is to first build an FET circuit that works well with a short sensing wire. Then the actual sensing circuit can be constructed and tried. Usually, it will pick up extraneous voltages from power line transients, radio stations, and a number of other sources. The situation can be helped somewhat by installing filtering and averaging circuits. The remainder of the problem is to make changes in the actual sensing circuit on a cut and try basis. Often it is more practical to use some other type of system. On the other hand, if the installer has the required time and patience, a good proximity system can be designed for many applications.

AN FET TOUCH ALARM

Fig. 6-8 and Table 6-3 describe the circuit of a very practical proximity alarm. The sensitivity has been reduced to the point where the intruder must touch or nearly touch the protective circuit to trigger the alarm.

A junction FET is used rather than a MOSFET, with the result that the circuit is much less sensitive and the protective transistor is no longer required.

The remainder of the circuit consists of an amplifier and a relay or trigger circuit to operate the alarm.

Table 6-3. Parts List for Fig. 6-8

Item	Description
Q1	FET, 2N5458 or equivalent
Q2	Transistor, silicon, npn, 2N3904 or equivalent
Q3	Transistor, silicon, pnp, 2N3906 or equivalent
R1	Potentiometer, 2.5 megohms, linear taper
R2	Resistor, 47K, ½ watt
R3	Resistor, 120K, ½ watt
R4	Resistor, 2.2K, ½ watt
R5	Resistor, 330 ohms, ½ watt
R6	Resistor, 470 ohms, ½ watt

AUDIO ALARMS

CHAPTER 7

The audio alarm is one that triggers an alarm in response to sounds in the protected area. It has the advantage that such an alarm can provide protection to a very large area. It has the disadvantage that it responds to any sound in the protected area, not only sounds made by an intruder. This isn't all bad, because in many areas excessive sound indicates a situation that requires attention. Thus if a normally quiet refrigerator or heater suddenly makes a lot of noise, it probably needs attention to prevent further damage.

The big problem is that noises from outside the protected premises may be loud enough to trip the alarm. Thunder often causes false alarms in audio systems.

The audio alarm is often used as a combination alarm and monitoring system. Sounds in the protected area will initiate the alarm and then the system can be used as a regular intercom. The listener can monitor the sound and make a judgment as to whether it indicates the presence of something that requires attention, or if the sound is caused by some extraneous noise.

Various types of sound cancellation systems have been built, but these often are unable to discriminate against outside noises arriving from different directions.

The sensor or detector of an audio alarm is usually a permanent magnet speaker of the type used in intercom systems. In fact, when building an audio alarm it is usually advisable to start with a regular intercom system that works well in a particular application. In this chapter, we will discuss methods of using the signal from an intercom system to operate one of the trigger circuits of Chapter 3.

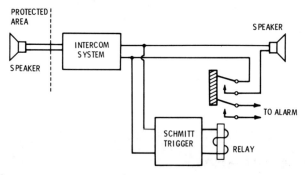

Figure 7-1.
Audio intrusion alarm using simple trigger circuit.

AN AUDIO TRIGGER CIRCUIT

Fig 7-1 shows a simple alarm in which a trigger circuit can be used with any audio amplifier, such as an intercom system, that may be already installed in the area where protection is desired. The sensor may be a permanent-magnet speaker of the type normally used in intercom systems or one of the vibration detectors described later. The intercom system in the figure is the audio amplifier. The remainder of the system is the trigger circuit, which is shown schematically in Fig. 7-2 (parts list in Table 7-1).

Transistors Q1 and Q2 form a circuit known as a Schmitt trigger. This circuit produces an output signal whenever the input signal

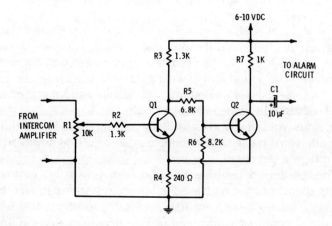

Figure 7-2.
Schmitt trigger circuit.

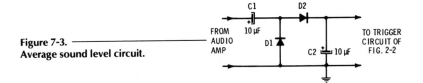

Figure 7-3.
Average sound level circuit.

reaches a preset level. With no input signal, transistor Q1 is cut off and Q2 is conducting. When the input signal reaches the level where Q1 starts to conduct, the voltage across emitter resistor R4 increases, tending to reduce the current in Q2. As the current in Q1 increases, the voltage at its collector decreases. This negative-going voltage is coupled through resistor R5 to the base of Q2, tending to reduce its collector current further. This action is regenerative and very quickly arrives at the point where Q1 is conducting and Q2 is cut off. As Q2 stops conducting, the voltage at its collector rises to the supply voltage. When the voltage at the input decreases, Q1 will again be cut off and Q2 will conduct.

In the circuit of Fig. 7-2, the rise in voltage at the collector of Q2 is coupled through capacitor C1 to an alarm circuit.

Note that this circuit will be triggered whenever the sound in the protected area exceeds a preset level. This is fine for applications where it is important to detect even a momentary increase in sound level, but it is impractical for most applications where it would be triggered by many harmless sounds.

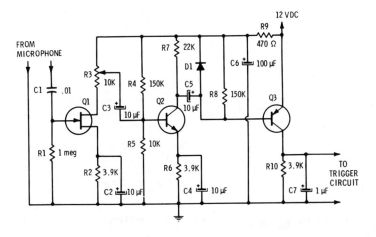

Figure 7-4.
Simple audio alarm.

A much more practical system would respond only to increases of the average sound level and would not cause false alarms on short, nonrepeated sounds. This can be accomplished by adding the circuit of Fig. 7-3 to that of Fig. 7-2. Here a rectifier and filter are included between the amplifier and the trigger circuit. The trigger circuit then responds to the average voltage developed across capacitor C2. The average sound level can be increased by increasing the value of C2. The optimum value for any particular application can be determined by experimentation. The parts list in Table 7-2 gives the range of typical values.

Table 7-1. Parts List for Fig. 7-2

Item	Description
C1	Capacitor, 10 μF, 25 volts, electrolytic
Q1, Q2	Transistor, silicon, npn
R1	Potentiometer, 10K
R2, R3	Resistor, 1.3K, ½ watt
R4	Resistor, 240 ohms, ½ watt
R5	Resistor, 6.8K, ½ watt
R6	Resistor, 8.2K, ½ watt
R7	Resistor, 1K, ½

A SIMPLE AUDIO ALARM

The two circuits described on the preceding pages use high-gain audio amplifiers and are very sensitive. In some applications, extreme sensitivity is neither necessary nor desirable. Often there is a need for a circuit that will respond to comparatively loud sounds only. Such a circuit is shown in Fig. 7-4 (parts list in Table 7-3). This circuit consists of a two-stage audio amplifier using an FET and a bipolar transistor. The amplifier is followed by a rectifier and an SCR trigger circuit.

Table 7-2. Parts List for Fig 7-3

Item	Description
C1	Capacitor, 10 μF, 25 volts, electrolytic
C2	Capacitor, 10-500 μF, electrolytic (see text)
D1	Diode, silicon, 1N4003
R1	Resistor, 10K, ½ watt

THE AUDIO ALARM AS AN ACCESSORY

Although audio alarms may be used effectively in many applications, there are many other situations where, because of the high am-.

bient noise, they are simply not very effective. They are not practical in areas where the background noise from appliances or motors is louder than the sound that would be made by an intruder. The alarm will either fail to sound as an intruder enters the protected area, or it will cause a large number of false alarms.

The audio alarm has one great advantage over other systems: it permits monitoring the protected area to evaluate what is going on before taking emergency action. Other alarms simply indicate that the alarm has been tripped, and there is no way to determine whether or not it is a false alarm. This advantage may be obtained with other alarms by adding an audio system as an accessory.

Table 7-3. Parts List for Fig. 7-4

Item	Description
C1	Capacitor, 0.01 μF
C2, C3, C4, C5	Capacitor, 10 μF, 35 volts, electrolytic
C6	Capacitor, 100 μF, 35 volts, electrolytic
C7	Capacitor, 1 μF, 35 volts, electrolytic
D1	Diode, silicon, 1N4003
Q1	Transistor, FET, 2N5458
Q2	Transistor, silicon, npn
Q3	Transistor, silicon, pnp
R1	Resistor, 1 megohm, ½ watt
R2, R6, R10	Resistor, 3.9K, ½ watt
R3	Potentiometer, 10K, linear taper
R4, R8	Resistor, 150K, ½ watt
R5	Resistor, 10K, ½ watt
R7	Resistor, 22K, ½ watt
R9	Resistor, 470 ohms, ½ watt

Fig. 7-5 shows an arrangement where an intrusion alarm is used to detect the actual intrusion. Once the alarm is triggered, the audio system is connected so that it is possible to monitor the protected area. The system can be made extremely sensitive. Since it is not used to trigger the actual alarm, there is no problem with high sensitivity causing false alarms.

A variation of this system that does not require anyone to be present at a monitoring point is shown in Fig. 7-6. Here an audio cassette recorder is connected so that it monitors the protected area whenever the alarm is triggered. With this arrangement it is possible to determine what happened in the protected area from the time the alarm was triggered until someone responded to it.

Most cassette recorders have a jack used with a push-to-talk switch to start the tape in motion (Fig. 7-7). In this case, it is not nec-

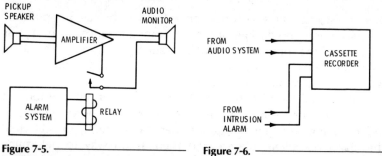

Figure 7-5.
Intrusion alarm connected to turn on audio monitor.

Figure 7-6.
Intrusion alarm turns on recorder.

essary to modify the recorder for the application. Connection from the alarm can be made through the push-to-talk jack.

The practicality of the circuit is limited by the amount of tape available in the cassette. Once the recorder is turned on, it will continue to record until the end of the tape is reached.

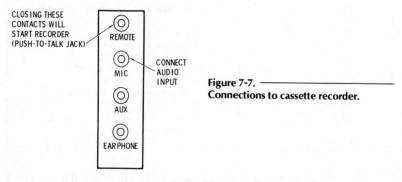

Figure 7-7.
Connections to cassette recorder.

SHOCK AND VIBRATION ALARMS

CHAPTER **8** —————————————————————————————

There are two different types of alarms that respond to shock or vibration. One is very similar to the audio alarms described in the preceding chapter. The only difference is that the sensing element is a vibration detector, rather than a microphone. Whereas a microphone converts sound signals in air to audio signals, a vibration detector converts the vibration of an object or surface into an audio signal. In either case the amplitude of the resulting audio signal is used to trigger an alarm.

Like the audio alarm, this type of vibration alarm can be monitored to obtain more information about what is actually happening in a protected area. If a vibration detector is attached to the wall of a building, it will respond to anything that causes the wall to vibrate, such as pounding to force a door open, or an attempt to cut through the wall. If the resulting signal is monitored on an ordinary speaker, the sound of the audio signal will give a clue as to what is causing the vibration. Such things as sawing, pounding, or drilling are easily distinguished.

The principal advantage of this type of vibration alarm is that it can be used in applications where an audio alarm would be impractical because of high ambient noise.

The other type of vibration alarm is very similar to an electromechanical alarm in that the sensing element is actually a switch. The principle is shown in Fig. 8-1. Here, a weight is suspended by springs. When the housing is vibrated, the weight will move in response to the vibration. If the amplitude of the vibration is great enough, the weight will touch the leaf of a microswitch causing it to open, just like the switch of an electromechanical alarm. Switches of

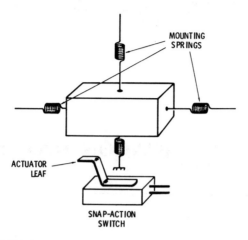

Figure 8-1.
Principle of vibration switch.

this type are commercially available. Such a detector can be used with any of the electromechanical systems described in Chapter 4.

This type of vibration alarm is useful in protecting against the theft of such things as tires, hubcaps, and accessories of automobiles. In such a theft the usual automobile alarm doesn't provide this type of protection because the automobile is not entered and the ignition is not tampered with.

AN AUDIO TYPE VIBRATION DETECTOR

Fig. 8-2 shows a vibration detector that can be constructed rather easily. The sensing element is a phonograph pickup and the needle rests on the protected surface. Inasmuch as audio quality is not a consideration, an inexpensive, high output pickup can be used. Pro-

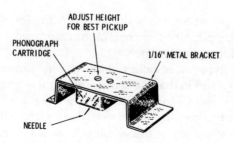

Figure 8-2.
Vibration sensor using phonograph cartridge.

vision is made to adjust the height of the pickup so that it can be set for maximum sensitivity. The rest of the system is one of the circuits for audio alarms described in Chapter 7.

APPLICATIONS

One very practical application of a vibration alarm is in protection of automobile accessories as mentioned previously. The other is in protecting an area against unorthodox methods of entry. The usual system protects an area against entry through doors and windows. In cases where the property being protected is very valuable, and the thief knows that an alarm system is installed on doors and windows, he might try to enter by cutting through a wall, floor, or roof. Although another type of alarm system may trip when the intruder actually reaches the protected area, it may not sound before damage is done to the floor, roof, or walls. A vibration detector can be used to supplement another type of system to prevent this destructive method of entry.

CLOSED-CIRCUIT TELEVISION

CHAPTER 9 ─────────────────────────────

Although the construction of a closed-circuit television system is beyond the scope of this book, commercially available cameras and monitors are priced so low that it is entirely practical to use them in home or small business protection systems.

There are two types of low-priced television cameras available. One type merely transmits a video signal that can be viewed on a monitor. The other type contains a miniature television transmitter and transmits a signal on a standard television channel so that it can be viewed on an ordinary television receiver without any modification.

The principal advantage of a television system is that it enables a person actually to see what is going on in the protected area. When combined with an audio system, it practically precludes responding to a false alarm. It is also possible to use the television signal to trigger an alarm; this will be explained later. Fig. 9-1 shows the components of a combined television and intercom system.

AN APARTMENT-HOUSE PROTECTION SYSTEM

The customary way of protecting apartment houses from unauthorized entry is to keep the front door locked. When a visitor pushes the doorbell button, the apartment dweller can talk to him on an intercom system and can unlock the door by means of a remote control system (electric door locks are discussed in Chapter 10). This system suffers from the limitation that positive identification is difficult with an audio system alone. When coupled with a television system, much more positive identification can be made.

Figure 9-1.
A combined closed-circuit television and intercom system. *(Courtesy Olson Electronics)*

A typical apartment house protection system is shown in Fig. 9-2. Here the camera is mounted so that it will pick up an image of anyone standing in the doorway. The output of the camera is on a television channel that is not used in the particular locality and is coupled directly into the master television antenna system of the apartment house.

Whenever the doorbell sounds, the apartment dweller may switch his television set to the proper channel and have a full view of the visitor on his television receiver. This, coupled with the ability to talk to the visitor on the intercom, will enable him to make positive identification.

A HOME TELEVISION PROTECTIVE SYSTEM

Fig. 9-3 shows a typical system that can be used to protect the grounds of a home surrounded by a fence. The television camera is located so that it will pick up an image of anyone at the gate. The television signal is coupled through a coaxial cable to a remote point in the house where the lock to the gate can be controlled.

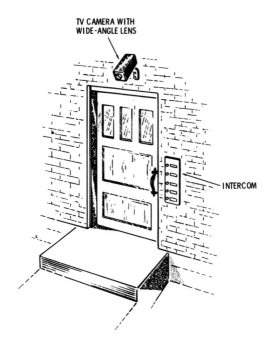

Figure 9-2.
A typical apartment-house protection system.

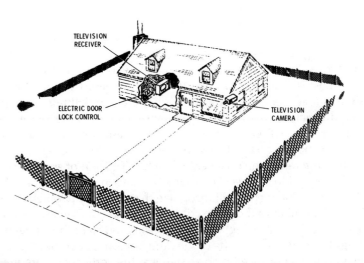

Figure 9-3.
A home closed-circuit television protection system.

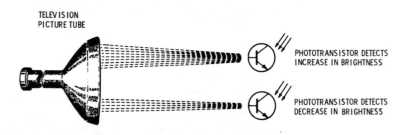

Figure 9-4.
Diagram of television light-sensitive circuit.

MISCELLANEOUS TELEVISION SECURITY SYSTEMS

Since a television system can be purchased at a comparatively low price and a regular television receiver can be used as a monitor, there are many applications where it is practical to add a television monitoring system to another type of intrusion alarm. For example, a television system may be used to control access to the back door of a store or business. Other applications include monitoring hallways in hotels or motels and parking lots.

A LIGHT-SENSITIVE TELEVISION ALARM

Although the most important use of a television monitoring system is to monitor an area that is protected by another type of intrusion alarm, it is possible to use the television system to trigger an alarm when the picture changes. A diagram of such a system is shown in Fig. 9-4. Here a small photoelectric light-sensitive circuit is mounted in a light-proof container so that the only light reaching the phototransistors comes from the television screen. The circuit is adjusted so that at a constant light level, the alarm will remain quiet. When the picture changes, the light level over part of the picture being monitored will either increase or decrease. In either case, the alarm will be triggered. The light sensor can be held against the screen of the television receiver by bracket (Fig. 9-5) or suction cup.

The television light-sensitive circuit is shown in Fig. 9-6 (parts list in Table 9-1). The reader will recognize this as simply two light-sensing circuits similar to those described in Chapter 5. One circuit is arranged to trigger if the light level increases and the other triggers if the light level decreases. The diode detector is included because the signal from the photocell is an ac signal at the vertical field frequency. The rectified signal is filtered by the 10-μF capacitor.

In using this circuit, it is important to have enough contrast to detect a change in the image, and yet keep the light level low enough so that the phototransistors will not be saturated. This is best accomplished by experimentation.

Table 9-1. Parts List for Fig. 9-6

Item	Description
C1, C2	Capacitor, 10 μF, 35 volts, electrolytic
C3	Capacitor, 10 μF, 10 volts, electrolytic
D1	Diode, silicon, 1N4003
Q1, Q3	Phototransistor, Motorola
Q2, Q4	Transistor, npn, Type 2N222
R1, R2	Potentiometer, 10-20K
R3	Resistor, 47K, ½ watt
R4	Resistor, 24K, ½ watt

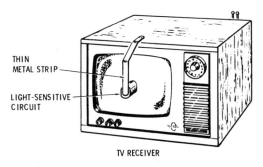

Figure 9-5.
Mounting television light-sensitive circuit to face of picture tube.

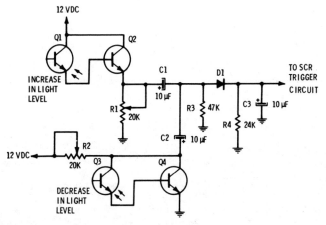

Figure 9-6.
Television light-sensitive circuit.

ACCESS CONTROL

CHAPTER 10 ───────────────────────────

One obviously necessary feature for any alarm system is some means of permitting an authorized person to enter the protected area without tripping the alarm, while preventing intruders from entering. Even the most sophisticated alarm is useless if an intruder can defeat the access control system. There are several methods of access control that can be used on an alarm. One obvious scheme is to simply use a hidden switch that will shut off the alarm. This rather naive approach seldom works because sooner or later someone will discover the location of the hidden switch.

KEY-OPERATED SWITCHES

An obvious type of access switch is the key-operated switch shown in Fig. 10-1. The switch is very similar to the ignition switch of an automobile.

Of course, the key-operated switch must be mounted in such a way that an intruder cannot easily gain access to the wires leading to it. Fig. 10-2 shows a typical arrangement with the switch mounted in a metal box.

Fig. 10-3 shows a protection system where both the protection circuit and the access lines are brought together to the access switch.

Figure 10-1. ───────────────────
Key-operated switch.

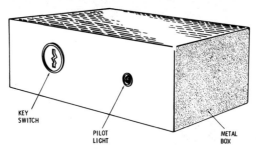

Figure 10-2.
Mounting the access switch.

The access switch opens the circuit in series with the SCR that triggers the alarm device. The protection circuit is connected to a normally closed snap-action switch which will open if someone tampers with the box containing the access switch.

This arrangement has an added advantage. If an intruder attempting to defeat the system cuts the protection circuit wires before he cuts the access circuit wires, he will trigger the alarm. If he shorts the wires together, he will either leave the alarm in operating condition or possibly trigger it.

COMBINATION SWITCHES

Another type of access control is the push-button combination switch shown in Fig. 10-4. With this arrangement the alarm will be disabled only when the buttons are pressed in the proper order. This type of access switch is very easy to build because a custom integrated circuit has been developed specifically for the purpose by LSI Computer Systems Inc. The circuit is designated as their Type LS7225 Digital Lock Circuit. It is quite versatile and can be used in many security applications.

Fig. 10-5 shows the pinout of the LS7225 and Fig. 10-6 shows a functional diagram. The operation of the circuit is described in the

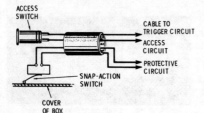

Figure 10-3.
Snap-action switch protects access box.

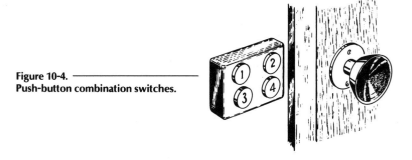

Figure 10-4.
Push-button combination switches.

following paragraphs in terms of what happens at the pins of the circuit. The power supply is connected to pins 3 and 6 with pin 3 positive and pin 6 grounded. When a series of positive signals is applied to pins 11 through 14 (Inputs I1, I2, I3, and I4) in the proper sequence, the SEQUENTIAL MEMORY is set. This causes the LOCK CONTROL OUTPUT (pin 8) and the MOMENTARY LOCK CONTROL OUPUT (pin 9) to go to a high state, and the LOCK INDICATOR OUTPUT (pin 7) to go to a low state. An external capacitor connected to pin 11 will determine the amount of time available to press all of the selected keys in the proper sequence.

The circuit operates from signals from four push-button switches. If desired, more than four push buttons can be used to confuse an intruder. The unused push buttons are connected to the UNSELECTED KEYS input (pin 10). If one of these unselected push buttons is pressed making pin 10 high the SEQUENTIAL DETECTOR will be reset and a pulse will appear at the TAMPER OUTPUT (pin 5). This pulse can be used to sound the main alarm, or some other alarm that indicates someone is tampering with the system.

The LOCK CONTROL OUTPUT (pin 8) will change state when the SEQUENTIAL MEMORY is set. This pin can be used to turn the alarm system on and off. The LOCK INDICATOR OUTPUT (pin 7) has an output which is the opposite of that of pin 8. It is used to drive an LED that indicates that the circuit is locked.

The MOMENTARY LOCK CONTROL OUTPUT (pin 9) goes high when pin 8 goes high, but it will revert to a low state when pin 11 goes to a low state. This time interval is controlled by a capacitor connected to pin 11. This output can be used when it is only desired to disable the alarm system long enough for someone to enter and restore the alarm to an active condition shortly thereafter.

The TAMPER OUTPUT (pin 5) gives a 15 μs pulse whenever pins 13 or 14 receive an out-of-sequence high signal from their respective

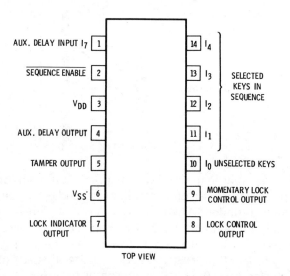

TOP VIEW

Figure 10-5.
Pinout diagram of type LS7225 Digital Lock Circuit. *(Courtesy LSI Computer Systems, Inc.)*

push buttons. The pulse will also appear whenever pin 10 receives a high signal.

The $\overline{\text{SEQUENCE ENABLE}}$ input (pin 2) enables the entire circuit. (The bar over the words "sequence enable" indicates that this pin responds to a low, rather than a high signal.) Thus, when this pin is forced to a high state, nothing will happen regardless of what push buttons are pressed. This input can be used with the TAMPER output and pins 1 and 4 to temporarily disable the system after it has been tampered with.

Fig. 10-7 shows the diagram of a push-button access control system using the Type LS7225. In this particular arrangement, the push buttons must be pressed in the order 4, 7, 2, 0, but this arrangement can be changed at will. There are two outputs. One relay operates for a few seconds only. The other relay stays energized until the circuit is reset either by temporarily disconnecting the power, or by pressing the push buttons in the proper sequence. When the alarm is on, the LED connected to pin 7 will light.

Capacitor C1 is normally not required for the circuit to operate properly, but it will provide noise immunity for locations where the ambient noise is high.

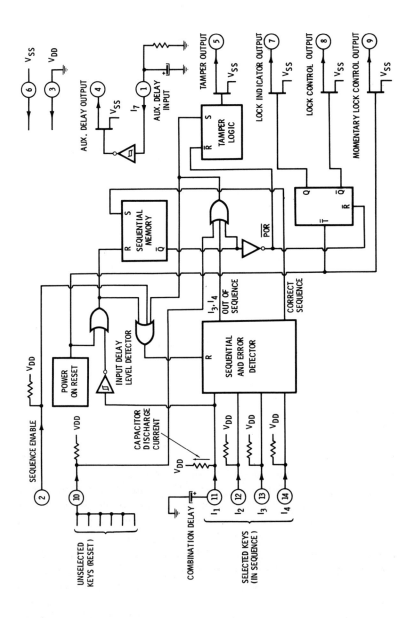

Figure 10-6.
Functional block diagram of type LS7225 Digital Lock Circuit. (*Courtesy LSI Computer Systems, Inc.*)

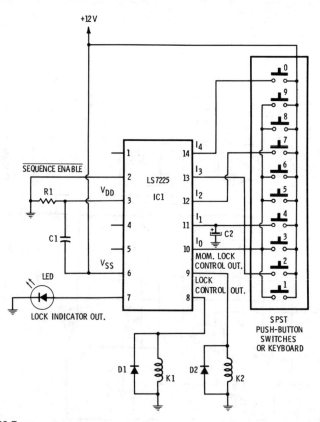

Figure 10-7.
Push-button access control.

The value of capacitor C2 determines the amount of time allowed to press the push buttons in the proper sequence. Making this capacitor smaller will reduce the allowable time.

The Type LS7225 is a very versatile circuit. With a little experimentation, many interesting variations can be developed. For example, in the circuit of Fig. 10-7 a hidden switch may be connected in series with the wire to pin 2. This will disable the circuit so that no amount of pressing the push buttons will accomplish anything. The circuit can also be used to operate an electric door lock. The door will then only unlock when the push buttons are pressed in the proper sequence.

The combination of the push-button switch can be easily modified with the arrangement in Fig. 10-8. Here the switch points from the circuit are brought to one side of a terminal board. The wires

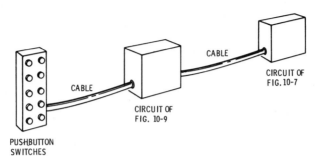

Figure 10-8.
Push-button combination switch arrangement.

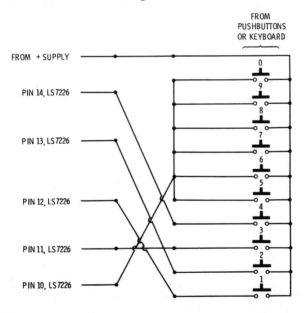

Figure 10-9.
Simple provision for changing combination easily.

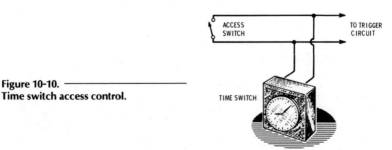

Figure 10-10.
Time switch access control.

from the push-button switches are brought to the other side of the board. The combination is set by wiring jumpers across the board, as shown in Fig. 10-9. Here the combination is set so that the relay will operate when button numbers 3, 1, 2, and 4 are pressed in that order.

TIME SWITCHES

A serious security problem with commercial establishments where items of great value are kept is that an intruder may actually force someone who has a key or knows the combination of an access switch to open the premises without triggering the alarm. This type of crime can be avoided with the circuit of Fig. 10-10. Here a time switch is connected with some other type of access switch. It is set so that the access switch will not operate except during certain hours. Thus, even if someone has a key, they cannot shut off the alarm except during normal business hours.

Table 10-1. Parts List for Fig. 10-7

Item	Description
C1	Capacitor, 0.05 μF, ceramic
C2	Capacitor, 4.7 μF, 35 volts, electrolytic
D1, D2	Diode, IN4003 or equivalent
LED	Any red LED will work
K1, K2	Relay, 12 volts, 500 Ω minimum
IC2	Type 7225 Digital Lock Circuit
	LSI Computer Systems, Inc.
	1235 Walt Whitman Road,
	Melville, NY 11747

Time switches also have other useful functions in an intrusion alarm. For example, in a home, the owner may want the system to sound a comparatively quiet alarm during the day and a rather loud alarm at night when he is asleep.

POWER SUPPLIES
FOR INTRUSION ALARMS

CHAPTER 11 ─────────────────────────────

Even the most carefully selected and constructed intrusion alarm is only as good as its power supply. Alarms will not operate without power. Even in localities where power failures are rare, the power line is subject to being cut by an intruder before the break-in. For these reasons, the conventional ac power supply that is used in most electronic equipment is rarely used in intrusion alarms. When an ac power supply *is* used, it is supplemented with standby battery supplies.

GENERAL CONSIDERATIONS

Since most intrusion alarms must operate properly without ac power, some type of battery must be used. The battery must be able to keep the trigger circuits operating in the event of a power failure until ac power can be restored. This is not a severe requirement with most of the circuits described in this book because most of them draw very little power. Another consideration is that the power supply must be able to operate the actual alarm device, such as a bell or siren, until it brings a response. This depends on the size of the alarm, which in turn depends on where the alarm is located.

There is a great temptation to build a power supply that is both battery and ac operated. The supply would operate from the power line except at the time of a power failure, when it would switch to the batteries. After power was restored, the batteries would be recharged. In general, this approach is discouraged because, although it is well within the state of the art, circuits of this type are beyond the scope of this book.

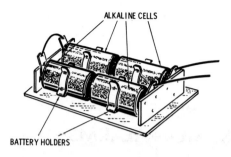

ALKALINE CELLS

BATTERY HOLDERS

Figure 11-1.
Typical battery power supply.

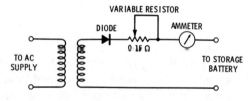

VARIABLE RESISTOR

DIODE AMMETER

TO AC SUPPLY

0 15 Ω

TO STORAGE BATTERY

Figure 11-2.
Battery-charger circuit.

The principal limitation of such an elaborate supply is that it is very difficult to meet all of the necessary requirements. The worst problem is usually triggering a false alarm when automatically switching between power-line and battery operation. Unless the supply is very carefully designed and tested, under certain conditions it will provide just enough of a transient to trip the alarm. Another consideration is to obtain the proper charging rate. Many automatic battery chargers generate enough noise to trigger an alarm while they are charging the batteries.

A final consideration in the selection of a power supply is that it must not transmit noise on the power line to the alarm system. In many localities, this is a severe problem.

BATTERY SUPPLIES

Probably the best power supply for an intrusion alarm is a battery. For small alarms, ordinary dry cells can be used. The number and type of cells depend on what is used for an alarm device. A small household alarm, such as a bell or electronic siren, will operate for several hours and can make plenty of noise when powered by the larger dry-cell batteries of the type used to power large spotlights. Inasmuch as an alarm should not draw any significant power unless there is an intrusion, the dry cells should last their normal shelf life. Fig. 11-1 shows a small dry-cell supply that will power a small alarm system.

The next step beyond a dry-cell supply is the use of rechargeable cells such as nickel-cadmium cells. These cells are more expensive and may not be worth the investment.

Larger systems can be powered by a motorcycle or automobile wet-cell storage battery. Such a battery will power even the loudest alarm for a long time.

The completely sealed lead-acid batteries that have come onto the market in the past few years are ideal for powering intrusion alarms. These batteries require almost no maintenance, except for occasionally checking the connections to the terminals to be sure that there is no corrosion. Special batteries of this type are designed to be floated across a regular power supply so that they will take over in the event of a power failure. Fig. 11-2 shows a circuit that can be used for recharging the batteries. The ratings of the diode, variable resistor, and ammeter depend on the size of the battery to be charged. Commercial battery chargers are also available at a reasonable price.

Of course, a battery supply for an alarm should not be accessible to an intruder. Otherwise, he could enter, trip the alarm, then immediately disable the supply. Fig. 11-3 shows a battery mounted in the same case with a charger. The case can be locked and firmly bolted to the floor with power leads brought out through the bottom or

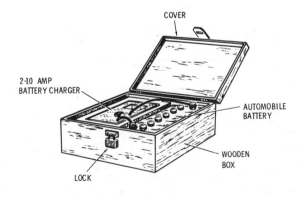

Figure 11-3.
Power supply using automobile battery and charger.

back of the case. An intruder could disable the supply, but in this event, it would take a great deal of time, and the alarm would be sounding while he was trying to disable it.

ALARM DEVICES
AND ACCESSORIES

CHAPTER 12 ⎯⎯⎯⎯⎯⎯⎯⎯⎯⎯⎯⎯⎯⎯⎯⎯

The purpose of an intrusion alarm is to get someone to respond in the event of an intrusion, to frighten an intruder away before he can steal anything, or, hopefully, to discourage an intruder before he actually attempts an intrusion. The device which attracts attention or frightens the intruder is the alarm device itself. It may be anything from a buzzer or bell to an electronic siren or one of the automatic telephone dialers described in Chapter 13.

ALARM DEVICES

In applications where it is necessary only to attract the attention of someone who is awake, a small buzzer or doorbell is an adequate alarm. Devices of this type are commercially available at any hardware store.

Where it is necessary to attract the attention of a neighbor, passerby, or policeman, a loud distinctive alarm is required. Many different types of bells are commercially available. Bells for intrusion alarms are designed and constructed so that they cannot easily be disabled by an intruder. In selecting a bell, a compromise is made between loudness of the bell and the amount of power required to operate it. An intrusion-alarm device must be able to operate when commercial power is off, and it must not use enough power to cause the batteries to run down before it attracts attention.

Sirens are becoming increasingly popular as intrusion-alarm devices. They not only attract attention, but they have a strong psychological effect on an intruder. A siren not only sounds like a police car, it also covers the sound of another siren; thus an intruder will realize that when a siren is operating, he cannot hear an approaching police car. Motor-driven sirens, such as the one in Fig. 12-1, and electronic sirens are commonly used.

AN ELECTRONIC SIREN

Fig. 12-2 shows a schematic diagram of an electronic siren that is simple, easy to construct, and very reliable. Any amount of audio power can be obtained by using this circuit with a regular public-address system.

The circuit consists of two relaxation oscillators using unijunction transistors. The stage containing Q1 is a low-frequency oscillator whose frequency is controlled by R3 and C1. The stage containing Q2 is a high-frequency oscillator whose frequency is controlled by R5, C2, and Q1. The output of Q2 is a high-pitched sound that can be altered by changing the value of R5. (Parts are listed in Table 12-1).

Table 12-1. Parts List for Fig. 12-2

Item	Description
C1	Capacitor, 10 μF, 35 volts, electrolytic
C2	Capacitor, 0.002 μF
C3	Capacitor, 0.15 μF
Q1, Q2	Transistor, unijunction, 2N2646
R1, R6	Resistor, 470 ohms, ½ watt
R2, R7	Resistor, 47 ohms, ½ watt
R3	Resistor, 47K, ½ watt
R4, R5	Resistor, 220K, ½ watt

By adjusting the two frequency controls (R3 and R5), the unit can be set to make a unique sound. If two or more alarms are in use in the same general area, they can be adjusted to make different sounds so that they can be readily identified.

The unit may be used for a small local alarm by connecting a 16-ohm speaker directly across the output without any further amplification.

FLASHING LIGHTS

Fig. 12-3 shows a flashing light of the type often used on police cars. It is powered by a 12-volt automobile battery. While this device is not adequate to attract attention and should not be used as the

Figure 12-1.
A motor-driven siren alarm.

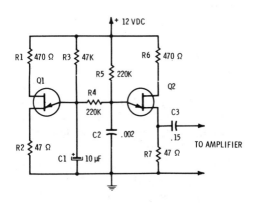

Figure 12-2.
An electronic siren circuit.

Figure 12-3.
A typical flashing light.

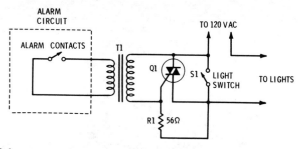

Figure 12-4.
Electronic light-switch circuit.

main alarm device, it is useful in identifying the exact location of an intrusion.

The flashing light has a strong psychological effect. When it is used with a loud electronic siren, the combined effect can be very unnerving to an intruder and may frighten him away before he can complete his mission.

BUILDING LIGHTS

A very effective deterrent to crime is plenty of light on the scene, particularly in heavily traveled areas. An intrusion alarm can be connected to turn on lights in an area whenever the alarm is triggered. This can be accomplished by connecting an electronic switch in parallel with the regular light switch.

Fig. 12-4 and Table 12-2 describe such an arrangement. Here, a triac is connected in parallel with the regular light switch. When the secondary of the transformer is open, there is very little current through its primary and hence very little voltage drop across resistor R1. The voltage at the gate of the triac will not be great enough to turn it on. However, when the intrusion alarm shorts the secondary of the transformer, a heavy primary current will occur, turning on the triac and all of the lights in the area. The triac will stop conducting when the alarm is reset.

Table 12-2. Parts List for Fig. 12-4

Item	Description
Q1	Triac (select to handle lighting load)
R1	Resistor, 56 ohms, 2 watts
S1	Switch, power, spst
T1	Transformer, 120 volts primary, 12.6 volts secondary

ACCESSORIES

In addition to the basic alarms described in various chapters of this book, there are many accessories that will enhance the effectiveness of the system. Some of these are described in the following paragraphs.

Time-Delay Circuit

There are many instances where it is advantageous to use a time-delay circuit in connection with an intrusion alarm. For example, if there is no response to an automobile alarm, there is little to be gained by letting the alarm run until it discharges the car battery. There are other cases where it is desirable for an alarm to reset after a few minutes. Some have been mentioned in earlier chapters.

Fig. 12-5 shows the circuit of an electronic time delay that can be set to any time from a few seconds to several minutes. (Table 12-3 is the parts list.) When the protective circuit is opened, SCR Q1 fires and causes a voltage drop across relay K1.

When Q1 fires, it effectively grounds the bottom of capacitor C1, which starts to charge through resistor R1. When the voltage across C1 reaches the firing potential of unijunction transistor Q2, it fires, causing a voltage drop across R3, which adds to the voltage across C2. This makes the anode of Q1 go somewhat negative, turning it off.

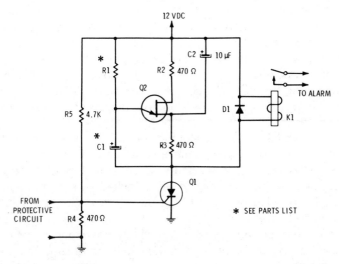

Figure 12-5.
Alarm with time-delay circuit.

Table 12-3. Parts List for Fig. 12-5

Item	Description
C1	Capacitor, 10-450 μF (see text), 35 volts, electrolytic
C2	Capacitor, 10 μF, 35 volts, electrolytic
D1	Diode, silicon, 1N4003
K1	Relay, 12 volts
Q1	SCR, 2N3228
Q2	Transistor, unijunction, 2N2646
R1	Resistor, 25K-2.5 megohm (see text), ½ watt
R2	Resistor, 470 ohms, ½ watt
R3, R4	Resistor, 470 ohms, ½ watt
R5	Resistor, 4700 ohms, ½ watt

This circuit will continue to operate for the preset amount of time, as long as the protective circuit is open. For example, if used with a door switch, it will operate for the set amount of time whenever the door is opened.

The time delay is controlled by the values of resistor R1 and capacitor C1. The time delay can be made adjustable by using a variable resistor for R1. In this case, a fixed resistor of about 25K should be inserted in series with the variable resistor to protect the unijunction transistor.

Electric Locks

Intrusion alarms are used not only to indicate an intrusion in a home or business when no one is there, but they are also used to alert a tenant to the presence of someone who may be an undesirable intruder. For example, an alarm may be used to detect the presence of anyone near the door of a house or an apartment. If a means of identifying the person—such as a television system or intercom system—is provided, an electric door lock may be used to allow a visitor to enter.

Fig. 12-6 shows a typical electric door lock. The lock may be operated by a push button, or it may be used in combination with one of the access switches described in Chapter 10. Fig. 12-7 shows an arrangement where a door can be unlocked by the push-button switch described in Chapter 10. The push-button circuit of Fig. 10-7 is connected so that when the proper combination of buttons is pressed, the door will unlock. An auxiliary power jack is provided so that the lock can be opened in the event of power failure of the main system.

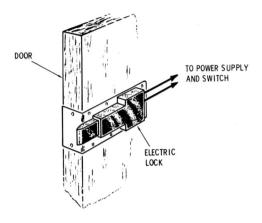

Figure 12-6.
A typical electric door lock.

Miniature Door Scope

An accessory that is useful with any home or apartment alarm is the miniature door scope shown in Fig. 12-8. It has a wide-angle lens and will enable anyone behind a door to identify a person outside. The purpose of an alarm is defeated if it is necessary to open the door to see who triggered the alarm.

Holdup Alarms

Any intrusion alarm may also be used as a holdup alarm by adding a switch to set off the alarm if a holdup is attempted. A holdup

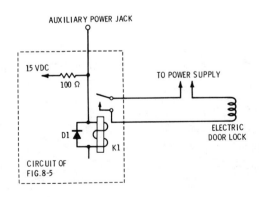

Figure 12-7.
A push-button door-lock circuit.

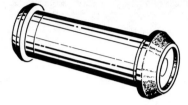

Figure 12-8.
A miniature door scope.

alarm switch must be well concealed and arranged so that it can be operated without a holdup man's attention. Otherwise, he might harm the person attempting to turn on the alarm.

In a store or business where there is a counter, the holdup switch can be arranged as shown in Fig. 12-9. Other arrangements will occur to the reader. An ingenious arrangement that can be used in a money drawer is shown in Fig. 12-10. Here a snap-action switch is arranged so that the plunger is held in one position by a dollar bill. When the bill is removed, the switch will operate and set off the alarm. In use, the bottom bill in a section of a cash drawer is clipped in the switch. In normal business, the bottom bill is never taken out of the drawer. However, a holdup man, not knowing this, will set off the alarm if he removes all of the money from the drawer.

An important but often overlooked application for a holdup alarm is in an automobile. Many crimes are committed when a motorist stops for a traffic signal or stop sign. If the automobile is equipped with an alarm that includes a loud siren, a hidden switch can be provided to set off the alarm. Then if a motorist is accosted when he stops for a red light, he can secretly flip the switch and set off the siren. The holdup man will often lose his nerve when attention is attracted to him in this way.

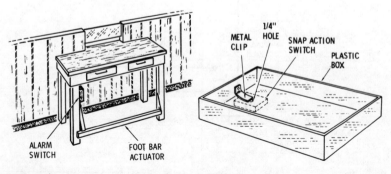

Figure 12-9.
Foot-operated switch permits using intrusion alarm as holdup alarm.

Figure 12-10.
Switch used to protect money drawer.

TELEPHONE ATTACHMENTS

CHAPTER 13 _____

One of the most serious limitations of an intrusion alarm is that the application may be such that it will not be heard. This is particularly true in rural areas or in business districts that are deserted at night.

Of course, a loud alarm always has some value, because an intruder can never be sure that the alarm will not be heard by someone. In addition, the noise will tend to be unnerving to him.

To assure that an alarm will be responded to in a remote area, it is necessary to transmit the alarm signal to some location where it will be heard and proper action will be taken. There are three common ways of doing this:

1. A radio link
2. A leased telephone line
3. An automatic telephone dialer

The advantages and limitations of each of these will be discussed in this chapter.

RADIO LINKS

The radio link is perhaps the most direct approach. As shown in Fig. 13-1, the alarm system installed in remote premises is connected to a radio transmitter which will send a signal to a monitoring point in the event of an intrusion. This arrangement is advantageous in that an intruder cannot defeat the system by cutting wires. However, a disadvantage is that the responder cannot know whether a component in the transmitter has failed.

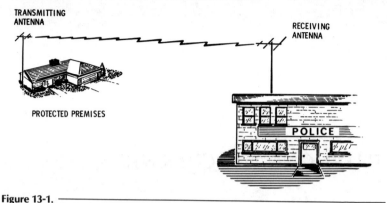

Figure 13-1.
Using a radio link to protect remote premises.

The nontechnical details of using a radio link are quite complex. First of all, a radio station license must be obtained from the FCC. This is not particularly easy, because with the current crowding of the spectrum, it must be proved that the license is in the best interest of the public. Furthermore, the transmitter used must be type accepted or type approved by the FCC.

LEASED TELEPHONE LINES

The next most direct system, shown in Fig. 13-2 uses a leased telephone line between the protected premises and the monitoring point. The principal disadvantage of this system is its cost, which depends on the distance the signal is to be transmitted.

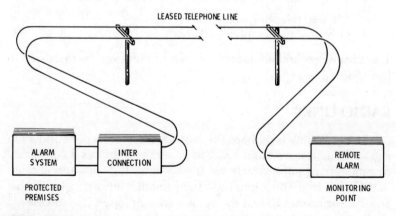

Figure 13-2.
Using leased telephone lines to protect remote premises.

Figure 13-3.
An automatic telephone dialer.

It would appear that the system could be defeated easily by simply cutting the telephone line. This can be avoided by an arrangement whereby the system transmits a signal at all times unless there is an intrusion. Then, if the line fails for any reason, an indication will be given at the monitoring point.

A system of this type can use any of the basic alarms described in this book, but the apparatus to be connected directly to the telephone line should be worked out in cooperation with the local telephone company.

AUTOMATIC TELEPHONE DIALERS

By far the most popular method of protecting remote premises is the use of a device that will automatically dial the police in the event of an intrusion. A typical unit of this type that is used with an audio alarm is shown in Fig. 13-3.

Superficially, it would appear that the automatic dialer is the ideal solution to the protection of all unoccupied premises. It is connected to the alarm system in such a way that in the event of an intrusion, the device will automatically dial the police department and report that an intrusion is in progress. The police can respond immediately and probably apprehend the intruder. On the surface, nothing could be simpler. In practice, however, automatic dialers have caused a number of problems. Most of these have been because of improper or unwise application.

The police departments of many large cities have reported that the alarm signals produced by automatic dialers in their areas have been over 90 percent false alarms. The natural result of this has been that the police have not taken reports from automatic dialers seriously. They see no point in risking lives and vehicles by rushing to answer what in all probability is a false alarm.

Another problem occurred in a city which reported that many different dialers had been tripped by restoration of power after a power failure. In one instance, after power had been restored, so many different dialers began calling the police department that the police telephone switchboard was completely jammed, and no telephone calls could get through to the police from any source.

Some of the minor troubles that have been encountered with dialers include: unintelligible recorded messages, dialers that would not disconnect after the message was sent, and interference and crosstalk on other telephone services.

As a result of these problems, some cities have passed ordinances governing the sale, installation, and use of automatic dialers that are arranged to call the police department in the event of an intrusion. In Dallas, Texas, which has such an ordinance, dealers who sell automatic dialers must be franchised by the city. Dialers that have been approved may be installed, but they must be set to dial a separate block of telephone numbers set aside solely for this purpose. Thus, if all of the dialers malfunction for some reason, such as interference from a thunderstorm, the regular police numbers will not be jammed.

USING AUTOMATIC DIALERS

In spite of the bad reputation that the automatic dialer has acquired in some areas, it can still be a very useful protection system if it is properly installed and used. The following guidelines are offered to help anyone who feels that the automatic dialer is the only practical solution to his protection problem.

1. Check the legal requirements of the local community. Be sure that the particular dialer selected is approved for use there.
2. Check with the police department concerning their requirements for dialers. Some police departments have a separate block of telephone numbers that must be used with automatic dialers. Others insist that the dialer must be set to call someone other than the police department who can relay the message.
3. Cooperate with the police department concerning the content of the message. They will probably want to use some type of code; if plain language is used, the intruder could defeat the system by calling and directing the police away from the actual intrusions.
4. Above all, be sure that the installation is reliable. Even if the police approve its use, they will soon lose confidence in it if they respond to a false alarm. It is advisable to have two separate intrusion alarms, operating on different principles and connected to the dialer in such a way that they both must be tripped before the dialer is actuated.
5. Coordinate the installation with the telephone company. They have experts to consult on any installation to assure that it will be reliable and will not interfere with other telephone services.

In situations where the police department is completely opposed to automatic dialers, these devices can still be used to summon someone other than the police in the event of an intrusion. For example, a home protection system may be set to dial one or more of the neighbors if an intrusion occurs when the owner is away.

In many instances where the cost is justified, the best way to protect remote premises is with a watchman. A local alarm or an automatic dialer may be used to alert him to an intrusion. In cases where one home or business cannot afford a watchman, several people can share the cost and have the watchman patrol the premises periodically.

PROTECTING THE AUTOMOBILE

CHAPTER **14** ──────────────────────

Automobile theft is by far the most common type of theft in the country today. In most larger cities, several thousand cars are stolen each year. For this reason and because automobile protection is somewhat unique, this entire chapter is devoted to the subject.

BASIC PRINCIPLES

An automobile is somewhat different than other objects from the point of view of an intrusion alarm. Whereas most objects can be stolen from premises that contain an alarm, an automobile must be taken with the intrusion alarm *in* it. For this reason, the thief must disable the alarm if he is to be successful.

There are two ways to minimize automobile theft. First of all, it should be difficult and time-consuming to accomplish the theft. The automobile should always be locked when it is unattended. Secondly, the alarm should be designed to make the thief nervous, so that he will give up before he successfully steals the car.

Since it is easy for the thief to open the hood of the car and "hot wire" it, any really secure automobile protection system will include a hood lock, such as the one shown in Fig. 14-1. A hood lock can be broken, but breaking it will take time and will discourage the thief.

AN IGNITION-PROTECTION SYSTEM

Fig. 14-2 shows a very simple, yet very effective, system for protecting the ignition of an automobile. It consists of a hidden grounding switch and a fuse in the line to the ignition key. When the car is running, the switch is opened and the only difference in the system

Figure 14-1.
A typical hood or trunk lock.

is that there is a fuse in the ignition line. The fuse is large enough to carry the normal ignition current so the engine will run properly.

When the automobile is to be protected, the hidden switch is closed. Now if anyone either turns the key or shorts it out, the full battery current passes through the fuse and the hidden switch, thus blowing the fuse. Now the engine will not start. Usually a thief will try another car if the first one that he picks cannot be started. If, instead of trying to jump the ignition key, the thief should try to "hot wire" the coil, his hot wire will become truly hot and will start to burn. This will probably confuse him enough so that he will abandon the attempt.

This simple system can also be used in connection with an alarm system that will attract attention while the thief is trying to determine why his usual ignition jumping procedures did not work.

A VOLTAGE-SENSITIVE ALARM SYSTEM

Fig. 14-3 shows a system that will initiate an alarm whenever anything is done to cause the voltage in the electrical system of the car

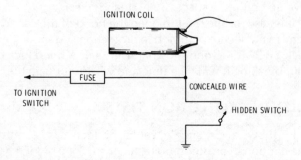

Figure 14-2.
Simple ignition-protection system.

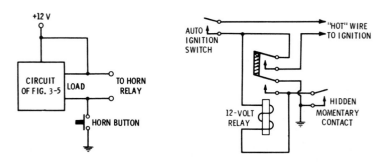

Figure 14-3.
Voltage-sensitive system.

Figure 14-4.
Automobile shaker alarm switch.

to drop. The trigger circuit is the voltage-sensitive circuit shown in Fig. 3-5 and described in Chapter 3. In this particular arrangement, the horn of the automobile is used as an alarm. Usually the current drawn by the dome light is enough to cause the voltage to drop enough to trigger the alarm. If not, the voltage drop incidental to starting the car will certainly trigger it.

The alarm can be reset by, of all things, momentarily depressing the horn button. This would hardly be expected as a way to stop the horn from blowing!

SELECTING TRIGGER AND ACCESS CIRCUITS

All of the trigger circuits described in Chapter 3 can be used in some way in an automobile. Any of the regular switches, such as door and light switches, can be used as intrusion-detection switches. The ignition line can also be used for this purpose.

The access switches described in Chapter 10 can be used both to turn on the alarm and as ignition switches. If circuits providing both turn-on delay and triggering delay are used, the owner can either enter or leave the car without tripping the alarm. The only restrictions are that he leave promptly after turning on the alarm and that he disable the alarm promptly on entering the car.

Protecting an automobile is difficult, but the amount of protection that can be provided is limited only by the experimenter's ingenuity.

Instead of using a trigger circuit to operate an alarm, it can be used to shut off a solenoid valve in the fuel line. With this system, the thief can steal the car, but he will not get very far with it. After the gasoline in the carburetor is used up, the engine will sputter and die. The thief will probably not realize that this is the result of a protec-

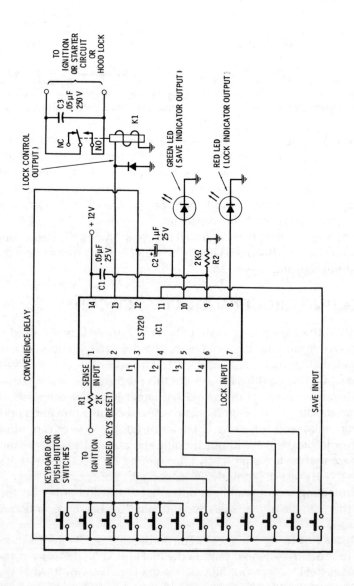

Figure 14-5.
Push-button ignition switch. *(Courtesy LSI Computer Systems, Inc.)*

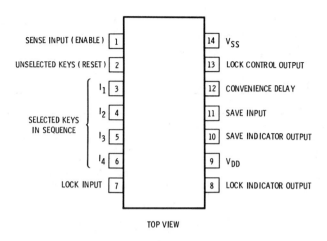

TOP VIEW

Figure 14-6.
Pinout of type LS7220 Digital Lock Circuit.

tive system, but will think that he unluckily stole a car that was either having trouble or was out of gas. In either case, he is likely to abandon the operation.

A PUSH-BUTTON IGNITION SWITCH

Fig. 14-5 shows the circuit of a push-button ignition switch that can be used in connection with a regular key-operated ignition switch. All of the circuitry is contained inside the Type LS7220 integrated circuit which is very similar to the Type LS7225 described in Chapter 10. Fig. 14-6 shows the pinout of the Type LS7220. When the regular ignition switch is turned on the red LED will light indicating that the circuit is is the locked state. When the keys are depressed in the proper sequence, the lock relay will be energized and the car may be started. This condition will be indicated by the red LED going out.

If the selected switches are pressed in an improper sequence or if any of the unselected switches is pressed, the circuit will be reset and the entire sequence must be repeated.

This circuit has a very interesting feature which makes it unecessary to reveal the proper combination to any one. Suppose, for example, it is desired that someone else drive the car as for valet parking. It is merely necessary to press the push buttons in the proper sequence to unlock the circuit, and then press the switch connected to the SAVE input. Now the green SAVE LED will light and the car can be started with the regular ignition switch. The combination lock will be out of the circuit. To disable the save circuit, merely press the push button connected to the LOCK INPUT, pin 7.

Table 14-1. Parts List for Fig. 14-5

Item	Description
C1	Capacitor, 0.05, 25 volts, ceramic
C2	Capacitor, 1μF, 25 volts, tantalum electrolytic
C3	Capacitor, 0.05 μF, 250 volts, ceramic
D1	Diode, type IN4003 or equivalent
K1	Relay, 12 volts, 500 Ω minimum
LEDS	Any red and green LED will work.
IC1	Type LS7220 Digital Lock Circuit LSI Computer Systems 1235 Walt Whitman Road Melville, NY 11747

SELECTING THE SYSTEM

CHAPTER 15 ─────────────────────────────────

With all of the different types of alarm systems available to the technician, it is often difficult to decide which system is best suited to a particular application. To some extent, almost any system can be adapted to almost any application; however, there is usually one system that will do a particular job better than any other system.

EVALUATING THE RISK

Before a system can be selected to protect anything, there must be a clear definition of just what is being protected against what threat and under what conditions. For example, there is a great difference in protecting a home against intruders when the resident is home and when he is away. When the resident is at home, he wants an alarm that will alert or awaken him when someone tries to enter the house. On the other hand, when he is away, he needs an alarm that will either scare the burglar away, call the police, or attract the attention of someone who will call the police.

There are several types of protection that may be required. There is *perimeter* protection that is designed to trigger an alarm whenever anyone enters the property being protected. There is *area* protection in which an alarm will sound whenever anyone enters a particular area of the premises, and there is *point* protection that will only trip the alarm when someone handles or approaches a particular object.

All intrusion alarms should be well concealed so that they cannot be defeated easily. However, the value of the protected property determines to some extent the effort a thief will exert to accomplish his mission. An ordinary home, where there are not many objects of high value, needs less protection than a palatial home that obviously contains many highly portable objects of great worth. In the latter case, it is advisable to use more than one intrusion alarm so that if an intruder defeats one, he may be caught by another. Multimode alarms are discussed later in this chapter.

OTHER SAFETY MEASURES

Perhaps the most important consideration in protecting any property is not to depend on the intrusion alarm to do any more than it is intended to do, which is simply to sound an alarm in the event of an intrusion. The alarm will not *prevent* the intrusion, but it may possibly scare off the intruder before he does any harm.

Other security measures are necessary with any intrusion alarm. Even the best alarm will have little effect if the intruder can complete a theft in a short period of time. He will simply enter, take what he wants, and leave before anyone has time to respond to the alarm that he has actuated. Therefore, doors should be solid and locks secure. Valuable objects should be secured mechanically so that it will take time to remove them. The philosophy should be that the alarm will attract attention to the presence of an intruder. Other measures, as many as are practical, should be taken to ensure that any theft will take as much time as possible. In this way, someone—owner or police—will have time to respond to the alarm before the theft has been committed.

ACTION REQUIRED

Selecting the alarm for a particular installation is also influenced by just what is intended to be accomplished. The homeowner is usually interested in protecting people and property. He wants an alarm that will attract as much attention to the intrusion as possible, so that the intruder will be frightened away. At the other extreme, the bank is very interested in apprehending the intruder. They may use a silent alarm to summon the police so that the burglar can be apprehended while the crime is still in progress. Most applications fall somewhere in between these two extremes.

THE IMPORTANCE OF SIMPLICITY

In this day of very sophisticated electronics, there is a great temptation for one to want the most "advanced" electronic intrusion alarm available. Advanced presumably means using very com-

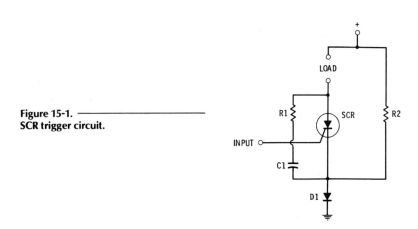

Figure 15-1.
SCR trigger circuit.

plex techniques. In security, this is a fallacy. The best security system is the one that will be the most dependable. It should always sound an alarm in the event of an intrusion, but never produce a false alarm.

Usually the most dependable alarm system is not the most complex or advanced system. It is the simplest possible system that will satisfy a particular requirement. This means that wherever practical, an electromechanical system should be used. Although it may seem old fashioned, with a little ingenuity, an electromechanical system can be adapted to almost all security requirements. It can also be made as foolproof as necessary.

When evaluating a system, it is helpful to analyze the system in terms of what actually triggers an alarm. Looking at the trigger circuit of Fig. 15-1, the alarm sounds whenever the SCR turns on. Of course, the system is designed so that an intrusion will cause the gate of the SCR to go positive, firing the SCR and thus causing the alarm to sound. Unfortunately, there are other things that can cause the SCR to turn on. If the anode voltage changes very suddenly, the SCR will turn on due to the dV/dt effect that was covered in Chapter 3.

The snubber circuit consisting of resistor R1 and capacitor C1 is designed to minimize the dV/dt effect, but it is possible that transients on the power line will get through the power supply to the anode of the SCR.

Another thing that will cause the SCR to turn on is noise pulses connected to the gate. The circuit consisting of diode D1 and resistor R2 is included to increase the noise immunity of the circuit.

Once a trigger circuit has been selected that is relatively immune to noise on the power line, a system should be selected that will not couple excessive noise into the gate circuit. With an electromechanical system, about the only way this can happen is by the protective circuit picking up stray pulses from a power line, nearby electrical machinery, or lightning surges. Protection against these things is relatively straightforward and is covered in Chapter 17.

Now let's look at a photoelectric system. The system is designed so that a change in the amount of light reaching the phototransistor, caused by the motion of an intruder, will initiate the alarm. Actually, any change in light, caused by almost anything, might also cause the alarm to trip. Thus we have one more external influence to worry about.

Furthermore, the photoelectric alarm usually has a high gain amplifier after the phototransistor. Naturally a high gain amplifier is much more susceptible to the influence of noise and extraneous signals than a simple trigger circuit. This gives us still another possible cause of false alarms.

Everything that we said about the photoelectric alarm will also apply to an audio alarm. In addition, the alarm can respond to very loud noises from sources well outside the protected area.

The most difficult of all is the proximity alarm. It is probably more sensitive to outside influences than any of the others. Thus it should be selected only when the other types are not practical.

This emphasis on simple alarms should not be construed to mean that more complex systems cannot be made dependable. They can. The point is that the more complex systems require a great deal more care in construction, application, and installation. Usually there are problems that the installer of a system does not anticipate. The simpler the system, the fewer of these problems are apt to arise.

TAILORING THE SYSTEM TO AN APPLICATION

Rarely will one of the systems described in this book be best suited, in its present form, to any particular application. It is possible, however, to tailor a system to a particular application. For example, an electromechanical system can be used to protect a motel room against theft of television receivers or other electrical appliances. Fig. 15-2 shows an intrusion switch built into an ordinary ac wall outlet. As long as the television is plugged into the outlet, the alarm will remain silent. When it is unplugged, the alarm will be triggered.

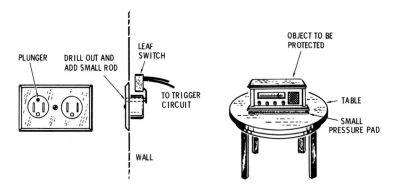

Figure 15-2.
Switch added to wall outlet protects tv set from theft.

Figure 15-3.
Using pressure pad to protect small objects.

In a store, it might be considered safe to allow customers to roam the area freely and handle most of the merchandise, but it might be unsafe if a medium-sized object, such as a radio or phonograph, is picked up. Fig. 15-3 shows how a pressure pad can be located under the object. The alarm will be triggered only if the object is lifted from the pad.

The specific adaptations that can be made are limited only by the ingenuity of the technician.

THE MULTIMODE SYSTEM

As noted previously, optimum protection of an area may best be accomplished by the use of more than one type of intrusion alarm. Fig. 15-4 shows a circuit that will initiate an alarm whenever one or more of the sensors connected to it is triggered. It also gives an indication of which sensor is active. (Table 15-1 is the parts list for this circuit.)

By the same token, it may be desirable to initiate an alarm only when two or more sensors, operating on different principles, are triggered. For example, when an automatic telephone dialer is arranged to dial the police in the event of an intrusion, it is highly desirable to minimize false alarms. The circuit of Fig. 15-5 combines the outputs from three different sensors. The circuit is actuated only when all of the sensors are triggered. (Table 15-2 is the parts list for this circuit.)

Fig. 15-6 shows a typical application of this arrangement. Here two electromechanical sensors and an audio sensor are used. If any

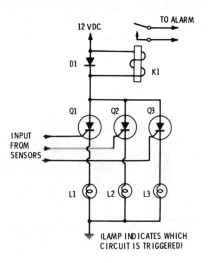

Figure 15-4.
Any of three sensors will trigger alarm.

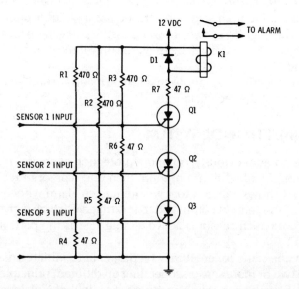

Figure 15-5.
A three-sensor trigger circuit.

one of the sensors should be triggered by something other than an intrusion, no alarm will be sounded, and there will be no false alarm. On the other hand, when all three sensors are triggered, there is a very strong probability that there is an intruder in the area.

Table 15-1. Parts List for Fig. 15-4

Item	Description
D1	Diode, silicon, 1N4003
K1	Relay, 6 volts
L1, L2, L3	Lamps, 6 volts
Q1, Q2, Q3	SCR

Table 15-2. Parts List for Fig. 15-5

Item	Description
D1	Diode, silicon, 1N4003
K1	Relay, 12 volts
Q1, Q2, Q3	SCR
R1, R2, R3	Resistor, 470 ohms, $\frac{1}{2}$ watt
R4, R5, R6, R7	Resistor, 47 ohms, $\frac{1}{2}$ watt

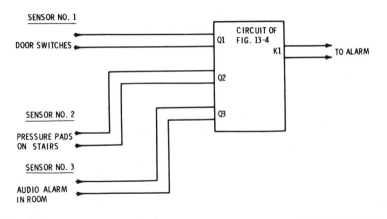

Figure 15-6.
Using three sensors to prevent false alarms.

MAKING IT WORK

CHAPTER **16** —————————————————————

Often the most difficult phase of constructing an electronic project is the debugging operation that is necessary to make the system work properly. There are four reasons why one of the circuits described in this book may not work properly after it is constructed and installed:

1. The components may not be correct. It is not always possible in all parts of the country to obtain the exact components specified, although every effort has been made to use commonly available components.

2. Some components may not meet specifications. This is particularly true when components are purchased from bargain sources. Unmarked or bargain components usually fail to meet one or more of the manufacturer's specifications. It is often better to use a first-line substitute component than to use a bargain component.

3. The circuit may not be wired properly. This is common, and it is necessary to check a circuit point by point when wiring errors are suspected.

4. There may be an ambient consideration that was unknown in the original design. All of the circuits in this book were tested over a limited temperature range and were subjected to the usual amount of electrical noise. There may be situations, however, where the ambient conditions are severe, and these should always be taken into consideration.

The biggest problem in dealing with distributors, particularly by mail, is that many of them will insist on a certain minimum order. Usually the total price of the components needed to build an alarm

(A) Charging circuit.   **(B) Voltage divider.**

Figure 16-1.
Effect of capacitor leakage.

system will exceed this minimum. However, procurement of a single component can be difficult.

There are many electronics supply houses that market their own brand of components. In many cases, they will provide a guide that will tell which of their components will replace the components specified here. Sometimes it may be necessary to actually try a component in a circuit to be sure that it will work.

The most tempting source of components is the advertisements in electronics magazines that offer standard parts at drastically reduced prices. Unfortunately, most of these components fail to meet one or more of the manufacturer's specifications. Even worse, the supplier can rarely tell you which of the specifications the component does not meet. Thus, there is the chance that a component will not operate at all in a particular circuit, or it may work initially but fail under conditions of extreme temperature.

On the good side, many suppliers of bargain components will guarantee the components that they sell. Thus, if a component fails to work in the circuit, they will replace it free of charge. This is, of course, a nuisance, but it will eventually lead to a workable system.

In each of the circuits in this book, every attempt has been made to avoid situations where the circuit depends heavily on the characteristics of the components. In most of the circuits, the values of the components may vary over a wide range without any ill effects. Probably the most critical component in any of the systems is the capacitor that is used to generate a time delay. These capacitors charge through resistors, and the equivalent leakage resistance of the capacitor acts with the series resistor to form a voltage divider. If the leakage resistance is too low, the capacitor voltage will never rise above a certain value. Fig. 16-1A shows a capacitor connected to charge through a 1-megohm resistor. If the leakage resistance of the capaci-

would look like the voltage divider shown in Fig. 16-1B. Here the capacitor would charge to one-half of the supply voltage and no higher.

Capacitor leakage tends to decrease (leakage resistance increases) after the capacitor has been charged a few times. This action is called *forming* an electrolytic capacitor. Therefore, if a capacitor is just sitting with no voltage applied to it, the leakage might be higher when voltage is first applied than it will be later. Thus it is a good idea to check time-delay circuits occasionally. Another factor to watch is that the leakage of electrolytic capacitors tends to increase with temperature.

CHECKING COMPONENTS

Most of the circuits described in this book are quite simple. It is probably just as easy to check a component in the circuit as it is to build a separate test circuit. In order to check a component properly, a high-impedance voltmeter such as a vtvm or solid-state voltmeter is required.

Fig. 16-2 shows a charging circuit connected to the emitter of a UJT. The principal problem with this circuit design is that the leakage of the capacitor might be high enough (leakage resistance low enough) so that the capacitor will never charge sufficiently to fire the UJT. When the circuit is working properly, the voltage across the capacitor will increase slowly at a rate determined by its capacitance and the value of the series resistance. When the voltage is great enough to fire the UJT, the voltage across the capacitor will drop and start building up again.

The impedance of the voltmeter used to measure the capacitor voltage will become a part of the circuit. For example, if the resistance in Fig. 16-2 happened to be 1 megohm and the voltage is being measured with a 20,000-ohms-per-volt meter, set on the 25-volt range, we would have the circuit of Fig. 16-3, where the 500K (500,000 ohms) is the resistance of the voltmeter. This circuit is obviously a voltage divider, and the voltage across the capacitor could never exceed 33% of the supply voltage. With this setup, a perfectly good capacitor could be discarded. If, on the other hand, a vtvm with a 10-megohm input resistance is used to make the measurement, we would still have a voltage divider, but the voltage across the capacitor could reach 90% of the supply voltage, and the test would be valid.

In fact, placing a high resistance of a few megohms in parallel with a component in a circuit should not interfere with the operation

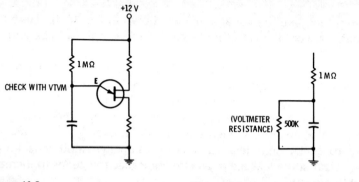

Figure 16-2.
UJT time-delay circuit.

Figure 16-3.
Effect of voltmeter loading.

of the circuit except in the transient state. If a capacitor is suddenly shunted by a resistor, the circuit might trip; but if a high-value resistor were placed across a component before the circuit was energized, the operation should not be impaired. This test often helps to locate critical areas in a circuit.

The pulse out of the UJT is so short that it usually cannot be measured with any degree of accuracy. A peak-reading voltmeter will indicate that the pulse is present, but may not give an accurate indication of its magnitude. If a UJT is suspected of not working, the resistor and capacitor in the time-constant circuit can be replaced with smaller values to give an output-pulse rate of about 1000 Hz. The output can then be observed on an oscilloscope.

Fig. 16-4 shows an arrangement for checking an SCR. Here the voltmeter is connected across a resistor in series with a variable supply connected to the gate. By applying Ohm's law, we can find out how much current is required to turn on the SCR. For example, the voltage drop across the 1000-ohm resistor in the figure will be one volt for each milliampere passing into the gate. If the current required to turn on the SCR is more than is available from the circuit,

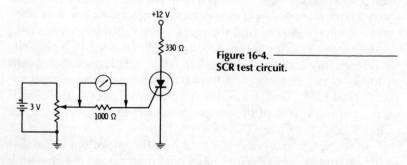

Figure 16-4.
SCR test circuit.

the system will not work. If an extremely small current is required to turn on the SCR, then extra precautions should be taken to prevent noise from triggering the SCR.

When the SCR is on, the voltage drop across it should be within the manufacturer's specifications. When the SCR is off, there should be no, or extremely little, voltage across the load.

An integrated-circuit timer can also be checked with a high-impedance voltmeter. The voltage rise across the timing capacitor, as well as the trigger and output voltages, can be observed. Usually a timer is either good or bad. Either the flip-flop will not change state at all, or it will operate properly.

Because of the close spacing between the pins of an integrated circuit, there frequently are wiring errors. A solder bridge between two of the pins is hard to notice, but it can keep the circuit from operating. It is also easy to miscount the pins, which results in a wiring error.

TEMPERATURE EXTREMES

The easiest way to handle temperatures that are too high or too low is to avoid them wherever possible. Electronic circuits are adversely affected by temperature extremes. Components tend to fail at elevated temperatures and batteries have low output at low temperatures. Often the temperature that a circuit will be subjected to can be changed considerably simply by relocating the circuit. The temperature gradient between the roof of a closed garage or attic and a spot a little lower can be several degrees. The same is true in a closed automobile. The temperature near the floor of the car may be several degrees lower than the temperature at the roof or under the dash.

If very high and very low temperatures will be encountered by a circuit that is used to protect valuable assets, the additional investment required to purchase components that meet military specifications may be worthwhile. MIL-SPEC components are much more expensive than commercial-grade components, but they will function properly under extreme conditions of temperature and vibration. When an application dictates the use of these high-grade components, the requirement should be discussed with the manufacturer or his distributor.

At any rate, there should be some assurance that a system will operate properly under the conditions it is likely to encounter. Temperature tests can be made using a household refrigerator and oven.

A portable thermometer is very helpful in making these tests. If a system fails a temperature test, measurements should be made to find out which component failed. That particular component can be replaced with one meeting tighter specifications.

The voltages used in alarm systems are low enough so as not to be hazardous except under extreme conditions. It is possible to pinpoint components that may fail at elevated temperatures by feeling them with a finger to be sure that they are not getting excessively warm under normal environmental conditions. Naturally, precautions should be taken to avoid electrical shock. If an ac-powered system is being tested, an isolation transformer should be used if any of the components are to be touched.

ELECTROMAGNETIC COMPATIBILITY

The expression *electromagnetic compatibility* refers to the situation where two or more devices that produce electromagnetic energy must exist in harmony with each other. The most common electromagnetic incompatibility occurs when a radio transmitter interferes with the reception on a tv set. An intrusion alarm may both radiate and be susceptible to electromagnetic energy. It may interfere with radio communication or it may be triggered by noise or radio signals.

It is common to neglect the interfering capability of an alarm system. The viewpoint is often adopted that if the alarm should happen to cause a little interference, it would attract additional attention to the intrusion. This is a shortsighted view, because, in general, if an alarm can radiate energy, it can also be triggered by spurious signals.

Susceptibility to interference might mean that an alarm would be triggered by interference, or that the interfering signal might cause it not to trigger in the event of an intrusion. Either defect is about as bad as the other. In any case, the alarm will not be reliable.

The kind of interference to which an alarm may be susceptible ranges from noise on the power line caused by electrical machinery or lightning to signals from a radio transmitter. The situation can be particularly difficult to detect when the offending transmitter happens to be a mobile unit that is only occasionally in the area. By the time that a technician is ready to start troubleshooting, the transmitter is no longer in the area.

Interfering signals may enter an alarm in any of three ways: The signal or noise may be present on the power line; it may be picked up by the protective circuits; or it may be picked up by the alarm circuits directly.

The first consideration in making an alarm immune to the effects of noise and spurious signals is to house it in a shielded enclosure. This usually means an aluminum box. The leads should be carefully brought into the box through filters when necessary.

Shielding is quite often misunderstood. Fig. 16-5A shows an alarm circuit that is completely housed in an aluminum box. The alarm is battery powered, so there are no power supply leads. The high side of the protective circuit is brought into the shielded enclosure through a small hole. The grounded side of the protective circuit is connected to the case, which is connected to a good ground. The alarm is located near some electrical machine that radiates appreciable noise. Superficially, it would appear that the system would be well protected from the noise. This is not true, however.

Noise contains high-frequency components, and high-frequency signals do not penetrate metal conductors very deeply because of *skin effect*. What actually happens to the circuit in Fig. 16-5A is shown in Fig. 16-5B. The high-frequency components induced in the case and the ground side of the protective circuit pass along the wire and the box as shown by the dashed lines. Because of the skin effect, the high-frequency energy does not pass through the bolt at the bottom of the case, but follows the surface of the case as shown. This means, in effect, that we have connected the case, which is supposed to be a shield, in series with one side of the protective circuit! Any noise induced in the case will appear across the circuits inside the case.

The proper way to connect a protective circuit to a shielded case is shown in Fig. 16-6. Here both sides of the protective circuit are brought inside the enclosure. The grounded side of the circuit is grounded to the inside of the case.

Interference conducted to the alarm system through the power line may be eliminated by installing a filter similar to that shown in Fig. 16-7 at the point where the power line enters the shielded enclosure. The coils in the filter can be made in the same way as the pulse transformer described in Fig. 2-15 of Chapter 2. The core is taken from a ferrite-rod antenna like those used in transistor radios. About 50 to 75 turns of wire are closely wound on the core. The wire size depends on the current that will be drawn. No. 16 wire will carry about 5 amperes. On a high-current circuit, larger wire must be used. In this case, it will be necessary to wind several layers to get the required number of turns. The layers may be separated with plastic electrician's tape.

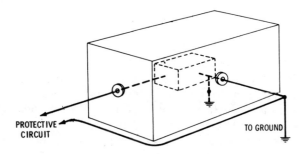

(A) Shielded.

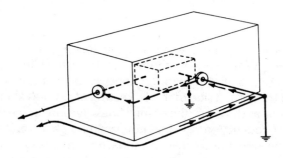

(B) Skin effect.

Figure 16-5.
Improper ground connection.

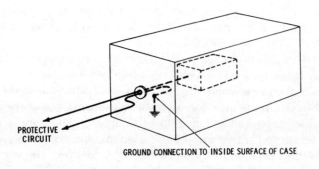

Figure 16-6.
Proper grounding.

The coils should not get hot when the alarm is drawing full current. If they do, larger wire is required.

Interference picked up by the protective circuit can be eliminated by using a filter at the point where the protective circuit enters the shielded enclosure, as shown in Fig. 16-8. Here the coil can be wound with very fine wire, because most protective circuits carry very little current. About 100 to 150 turns wound on a ferrite core will usually do the job nicely.

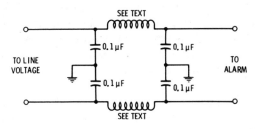

Figure 16-7.
Power-line filter.

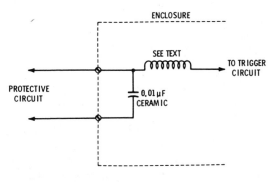

Figure 16-8.
Protective-circuit filtering.

SUMMARY

If an intrusion alarm does not work after it is constructed, the following procedure should isolate the trouble:

1. Reread the description of the circuit. Be certain that you know how it is supposed to work. This will often give a clue as to what is wrong.
2. Check the components to be sure that they are correct. Review the material in this chapter on tests.

3. Check the wiring. This is the most common reason that circuits do not work.
4. Check to see if improper operation is due to outside influences such as extreme temperatures or electrical noise.

SAFETY CONSIDERATIONS

An intrusion alarm is like any other piece of electronic equipment as far as safety is concerned. Protective circuits, in particular, are apt to be touched by children. The design must be such that no lethal current can be drawn from any exposed parts of the system. All ac-powered systems should use a power transformer. The leakage current of the transformer should be checked to be certain that it is not excessive.

All metal parts except the conductors should be grounded. In an ac-powered system, a metal case, if used, should be connected to the third wire of a three-wire plug.

Like any other electronic system, an intrusion alarm can develop excessive heat if a short circuit occurs. The power leads should be fused if there is any source that can provide excessive current.

ELIMINATING FALSE ALARMS

CHAPTER **17** ──────────────────────────────

We have stressed earlier that the false alarm is probably the most annoying type of fault that can occur in an intrusion alarm. It is for this reason that an intrusion alarm must use carefully selected components as well as careful construction.

Unfortunately, even a perfect intrusion alarm can be subject to false alarms that are caused not by failure of internal components, but by external influences. Finding the cause of a false alarm can be a very time consuming and tedious process. Usually whatever caused a false alarm is long gone by the time that a technician tries to locate it. There are, however, a few steps that will make the process easier.

Before we describe the procedures that will help to locate the causes of false alarms, it is important to stress that every false alarm should be investigated. Everyone very quickly learns to distrust an alarm system that has a history of false alarms. For this reason a prospective burglar may find ways to cause false alarms. There are cases where burglaries have taken place with impunity because an alarm hasn't even been turned on to avoid the inconvenience of false alarms. A trick that is used by prospective burglars is to trip an alarm so that they can measure the response time. If enough time elapses between the time an alarm is tripped and when someone responds to it, a burglar may find that he can enter the premises, trip the alarm, take what he wants, and get away before anyone arrives on the scene.

Thus a false alarm may be caused by someone who is planning to enter the premises at some later time. If no other cause can be found, this possibility should be seriously considered.

EXTERNAL INFLUENCES

There are several ways in which an external influence can enter a system. As shown in Fig. 17-1 the influence may enter through the sensors, the wiring to the sensor, the power line, and the circuits of the system.

The sensor of an alarm system doesn't respond directly to the presence of an intruder. It responds to something that an intruder does. In an electromechanical system, the system responds to a change in the voltage or current in the protective circuit. Ideally, this change in voltage will be caused by the intruder opening a protective circuit, but if another influence changes the voltage in the protective circuit sufficiently, the alarm will be tripped.

A photoelectric system responds to a change in light or infrared level. Ideally, this change will only be caused by some action of an intruder, but again an external influence may also cause such a change. The same considerations apply to audio, vibration, and proximity alarms.

Influences entering through the wiring, the power line, and the circuits of the system are electromagnetic in nature. These may be caused by surges on the power lines, lightning surges, or radiation from other devices.

AMBIENT CONDITIONS

One of the best approaches to finding the cause of a false alarm is to learn just what the alarm system might have been subjected to when the false alarm occurred. When an alarm is installed in a home, the owner may have first-hand knowledge of these conditions. For example, he will know whether or not there was a thunderstorm or a power outage at the time that the false alarm occurred. Usually, the

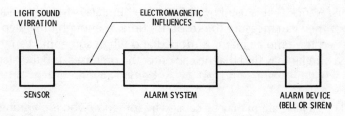

Figure 17-1.
Outside influences that can trigger false alarms.

technician investigating a false alarm will have little first-hand knowledge of the ambient conditions when the false alarm occurred. There are, however, many sources of information that will help to reconstruct the conditions.

In most locations it is comparatively easy to determine the temperature, humidity, and the presence of storms at the time of occurrence of a false alarm.

The nearest office of the National Weather Service can usually supply a great deal of information on weather conditions. Newspapers, radio stations, and the local police departments can also help.

Unusual conditions on the power lines can also be determined with comparative ease. Most utilities are quite cooperative in supplying information about power line voltage, transients, and surges when they know the purpose of the inquiry.

Transients on a power line that are caused locally by some electrical device are not as easy to trace. Often such a disturbance that is great enough to cause problems locally is not great enough to show up on the instrumentation of the utility company.

A good source of general information that might be helpful is the local law enforcement agencies. Usually police departments dislike false alarms in security systems because they interfere with normal law enforcement procedures. When the reason for the inquiry is explained in detail, they can often provide a great deal of significant information. The key is to ask the proper questions. An audio alarm might respond to the sound from the siren of a police car, ambulance, or fire engine. The police will generally know the time of occurrence of these things. The vibration from an explosion, or accident, might also cause a false alarm. Again, the police are usually aware of these things.

In general, as much information as possible should be collected about all of the ambient conditions at the time that the false alarm occurred. Sometimes a seemingly unrelated event can shed light on what actually caused the alarm.

TRACING THE EFFECT OF AMBIENT INFLUENCES

Once the ambient conditions related to a false alarm are determined the next step is to attempt to see if there is any correlation with the false alarm. If something happened at the exact time as the false alarm there is usually a correlation. Suppose, for example, that there was a power outage and a false alarm occurred when the power was restored. There is usually a correlation. Even if the alarm itself will withstand the application and removal of power without tripping, some other device might cause interference when power is restored.

If an emergency vehicle passed the alarm at the time a false alarm occurred, the actual cause might be noise or vibration, or radiation from a two-way radio.

Unfortunately, there are many false alarms where the technician cannot establish the exact time of occurrence. Trying to correlate the false alarm with an external influence can then be much more difficult. For example, a power outage might have occurred the same evening, but it may have not occurred at the same time. In this situation, the power line disturbance is suspected, but there is no definite correlation.

ELECTROMAGNETIC INFLUENCES

By far the most common cause of false alarms in security systems is some sort of electromagnetic influence. This can range from a surge on a power line, or an impulse due to a lightning flash, to radiation from some device such as a two-way radio. There is a very subtle way in which radiation from something like a tv or broadcast station can contribute to false alarms without causing them directly. A nearby powerful radio station may induce a substantial voltage in the wiring of an intrusion alarm. Although this voltage may be substantial, it might not be great enough to trip the alarm. The alarm can trip, however, when a relatively small additional surge is induced in the wiring.

This situation is shown in Fig. 17-2. In Fig. 17-2A, the voltage on the wiring without interference is shown. In Fig. 17-2B, there is a spike induced on the alarm. Note that the spike isn't great enough to trip the alarm. In Fig. 17-2C we have the same spike, but now it is superimposed on a signal from a broadcast station. The total signal is now great enough to trip the alarm, although neither signal alone would do it.

THE POWER LINE

There are two ways in which the power line can carry interfering signals into an alarm system. These are shown in Fig. 17-3. The most obvious way in which a power line can carry surges into an alarm is through the power supply, as shown in Fig. 17-3A. The other, more subtle way is shown in Fig. 17-3B. Here the power line does not carry

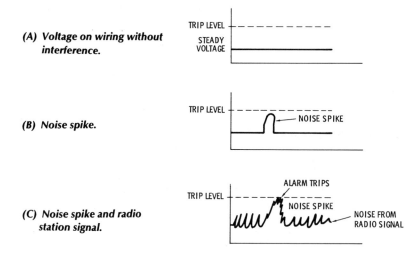

(A) Voltage on wiring without interference.

(B) Noise spike.

(C) Noise spike and radio station signal.

Figure 17-2.
Two influences combining to produce a false alarm.

interfering signals directly into the system. Rather, it carries the interference to the proximity of the wiring of the system, such as a line from a sensor. Inasmuch as the power line may be behind a partition, its presence is often unsuspected.

Finding interfering signals on a power line isn't straightforward. The presence of the 60-Hz line voltage on the line tends to mask any other signals that may be present. One way to get a good look at what is actually present on a power line is to use an oscilloscope in connection with a 60-Hz reject filter.

Table 17-1. Parts List For Fig. 17-4

Item	Description
C1, C2	5 µF, 400 volts*
C3	10 µF, 400 volts*
R1, R2	1000-Ω Pot, set to 530 Ω
R3	1000-Ω Pot, set to 215 Ω and adjust for maximum attenuation.

* Can be made up of 1- or 2-µF capacitors in parallel.

The Twin-T circuit of Fig. 17-4 will permit looking at a power line for noise without seeing the normal 60-Hz power line voltage. The Twin-T circuit is actually quite remarkable. It will provide infinite rejection of a single frequency if it is properly balanced. The circuit val-

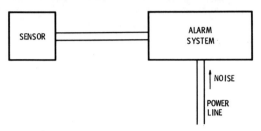

(A) Directly through power supply.

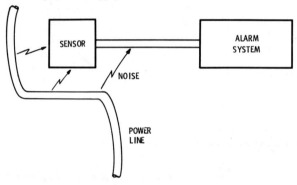

(B) Radiated from power line.

Figure 17-3.
Noise from power line.

ues required to obtain infinite rejection are given by the equations in Fig. 17-4. Unfortunately, these equations usually do not lead to standard component values. A satisfactory network for rejecting 60 Hz can be made with the capacitor values shown and using pots for the three resistors. The parts list is given in Table 17-1.

When setting up the circuit, adjust R1 and R2 close to 530 ohms with an ohmmeter and then leave them alone. Adjust R3 to about 215 ohms. Later this pot can be adjusted for maximum rejection.

Fig. 17-5 shows the Twin-T circuit connected in the input of an oscilloscope. With this arrangement, care must be taken to be sure that the grounded side of the power line and the ground input of the oscilloscope are both connected to the bottom line of the Twin-T network. Otherwise the full 120-volt line voltage may be between the case of the oscilloscope and ground, presenting an electrical shock hazard.

Once the circuit is connected as shown in Fig. 17-5, pot R3 can be adjusted to minimize the line voltage display on the oscilloscope.

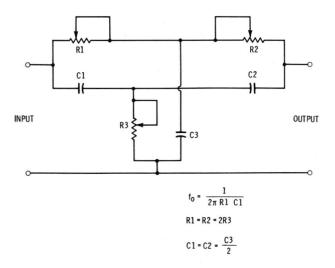

$$f_0 = \frac{1}{2\pi\, R1\ C1}$$

$$R1 = R2 = 2R3$$

$$C1 = C2 = \frac{C3}{2}$$

Figure 17-4.
Twin-T rejection network.

As the line voltage display becomes smaller, the sensitivity of the oscilloscope can be increased.

When the circuit is adjusted properly, very little of the 60-Hz line voltage will be seen on the screen of the oscilloscope. In most locations the oscilloscope will then show all sorts of spurious signals and surges. The display will show such things as signals from nearby radio stations as well as ac control signals for such things as traffic lights. Often it is obvious that the power line must be cleaned up before an alarm can be operated without numerous false alarms.

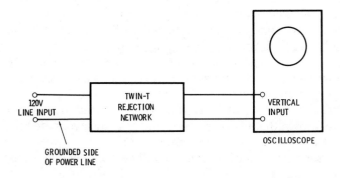

Figure 17-5.
Using a rejection network to find interfering signals with an oscilloscope.

Sometimes excessive noise on a power line can be found by simply bringing a small transistor radio close to the line. This technique is often useful in locating noisy power lines that run within partitions. If one area of a wall is found to induce more noise in the radio than other areas, there is probably a power line in the noisy area. Such a place on a wall is not a good place to run the wires from the sensor of an intrusion alarm.

CLEANING UP THE POWER LINE

It is usually a quite straightforward task to keep interference from entering the alarm through the power supply, although severe cases may require patience. There are two separate types of troublesome signals that may be present on the power line. The first is what we might call surges. These are spikes of high voltage that are present on the line. These spikes may be caused by lightning, or by switching transients that occur when something goes wrong in the power distribution system. It is not uncommon to find transients of over 1000 volts on a 120-volt power line. In most systems these transients are of such short duration that they don't cause any problems. In an alarm system, however, they can easily cause a false alarm.

Filters are usually not very effective in suppressing surges. Although the filter will attenuate the surge considerably, it will often have enough energy to produce a signal of troublesome level at the output of the filter. The best approach is to use some kind of surge suppressor. The GE-MOV metal oxide varistor made by the General Electric Company usually works quite well.

Fig. 17-6 shows a functional sketch of a MOV transient suppressor. It consists of fine grain polycrystalline semiconductor crystals that are fabricated with a ceramic material. The device operates much like two zener diodes connected back to back. Under normal line voltage conditions it has a very high impedance. However, when it is exposed to a high voltage transient, its impedance will drop to a very low level, thus clamping the peak voltage to a safe level. The energy of the transient will be safely dissipated in the device.

Fig. 17-7 shows a MOV suppressor connected in an alarm system. As shown the device is connected directly across the line, just inside the shielded enclosure of the system. Many times the addition of such a transient suppressor will completely eliminate false alarms.

The other type of interference that can be carried into a security system is a spurious signal. Its level might not be very high; it would

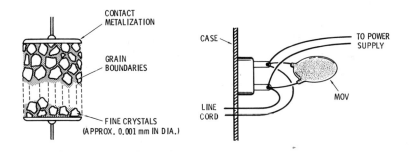

Figure 17-6.
GE-MOV™ transient suppressor. *(Courtesy General Electric Co.)*

Figure 17-7.
An MOV surge arrestor connected across the power line.

be much lower than the line voltage. However, if its frequency is such that it will influence the circuits of an alarm system, it can cause false alarms. For example, a spurious signal having a frequency in the audio or video range may raise havoc with an audio alarm.

This type of interference is controlled by filtering and shielding of the alarm system. It is usually a good idea to enclose as much as possible of the circuitry of an alarm system in a good shielded enclosure. When selecting the shield and connecting it, the skin effect considerations mentioned in Chapter 16 must be taken into consideration.

Interference suppression filters can be thought of as impedance *mismatching* devices. As shown in Fig. 17-8, the filter is connected between the power line and the input to the power supply of the system. The purpose of the filter is to pass the 60-Hz line voltage without attenuation, while attenuating signals of higher frequencies. The attenuation is accomplished by making the power-line side of the filter look like a high impedance to the interfering frequencies. Thus a low-pass filter is required.

The arrangement of Fig. 17-9A would work fine if all of the interference were between the two conductors of the power line. This type of interference is usually only a problem when the source of the interference is reasonably close to the alarm system. Usually if the source of interference is more than a few hundred feet away this type of interfering signal will be attenuated in the power line and the various devices that are connected to it.

The most troublesome type of interference is called common-mode interference and it is carried on both of the power lines with respect to ground. This type of interference can travel long distances.

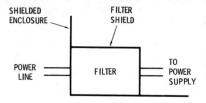

Figure 17-8.
An interference filter, shielded and connected between the line and power supply.

Table 17-2. Parts List for Fig. 17-9

Item	Description
C	10 µF, 400 volts
L	100 Turns*, ½" in diameter

Wire size great enough to handle full power supply current.

To eliminate both types of interference, a balanced filter of the type shown in Fig. 17-9B can be used. The values shown in Table 17-2 are tentative. It is not possible to specify the component values exactly, because the exact impedances of both the power line and the power supply of the system at any particular interfering frequency will differ in different applications.

DIRECT PICKUP

Once the system has been made immune to interference or surges entering through the power line, the remaining problem is direct pickup of interference from the wiring of the system or the circuits. The circuit problem can be minimized with a good shield.

The pickup in leads from the sensors or leads to an alarm device such as a bell presents a difficult problem. The most difficult part is trying to find out where the signal is originating.

We have mentioned the fact that power line wiring can carry surges close to the wiring of the alarm system. Other wiring can

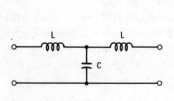

(A) Filtering between conductors.

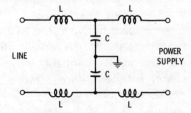

(B) Filtering between each conductor and ground.

Figure 17-9.
Power line filters.

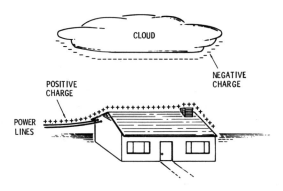

Figure 17-10.
Charge buildup during a storm.

cause the same problem. For example, lightning surges on a telephone line can find their way into alarm system wiring that is close
by.

Any type of conductor can cause surge problems. In fact most of
the problems caused by lightning are not caused by direct radiation
or conduction from the lightning stroke, but by surge currents that
the lightning discharge causes in other conductors. Furthermore,
many of the surges associated with thunderstorms are not associated
by actual lightning strokes but by static charges that are built up by
the storm. Fig. 17-10 shows a sketch of what happens in a thunderstorm. Essentially the clouds carry a large electric charge. As they pass
by an opposite charge is induced in the earth and in objects on the
earth. When the potential difference between the clouds and the
earth becomes great enough the air ionizes and a lightning stroke occurs. In this instance there is obviously a large current flowing along
the path of the stroke and a large magnetic field is produced. This
field will cause currents to flow in all other conducting devices that
are close by.

What is generally not realized is that there are many surges in a
thunder shower that are not associated with direct lightning hits. The
induced charges can cause surges even when there isn't a great
enough potential difference to cause a lightning stroke. For example,
large surge currents have been measured in radio antennas during
storms even when the nearest lightning was over a mile away.

Thus we can expect large surge currents to flow during a thunderstorm in such conductors as ground wires from lightning rods,
drain pipes, and even metal reinforcements in concrete buildings.

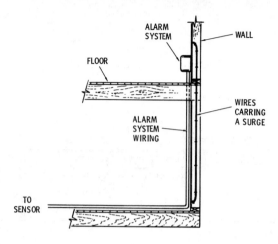

Figure 17-11.
Induced surges from other wiring.

Fig. 17-11 shows a typical situation where static discharges might cause problems. Here the security system is located on an upper floor of a building. It is arranged to protect, in addition to other things, the front door on the first floor of the building. Wiring from the first-floor sensor is run vertically to the alarm system. This vertical run may be inadvertently close to some other conductor that will carry high currents from static discharges. Whether or not the surges will cause a false alarm depends on the details of the installation and the magnitude of the surge current. Such a system might cause false alarms in some thunderstorms and not in others.

Another source of very large static charges that is not generally appreciated is in the wind that blows well in advance of a storm. Although there may be no thunderstorm activity in the immediate area, large currents may flow in vertical conductors. Broadcast engineers are particularly aware of this situation because of the arcing that occurs across the insulators in antenna guy wires when there is no obvious storm activity in the area.

What has been said about surges originating with storms also applies to other very large surges that might be due to such things as a

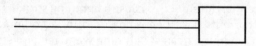

Figure 17-12.
Equal voltages are induced in symmetrical conductors.

short circuit in electrical equipment. The short currents that flow before a breaker trips may induce currents in other conductors that run close to the wiring of an alarm system.

MINIMIZING PICKUP

If two wires such as those shown in Fig. 17-12 are completely symmetrical and are exposed to an ambient electromagnetic field, equal voltages will be induced in both of the wires. If one of the wires is grounded at the alarm system, the voltages between the two wires will be the same and at ground potential. No voltage will be carried into the system. Of course, in practice, it is very difficult to keep both wires completely symmetrical throughout their entire length. For example, if the sensor connected at the far end of the wires happens to be a window foil as in Fig. 17-13, the two paths will be quite unsymmetrical at this point. Even so, if the wiring is kept symmetrical over as much of its length as possible, pickup will be minimized.

The susceptibility of wiring and a sensor to spurious signals can be judged by disconnecting the wires from the alarm system (but not from the sensor) and looking into them with an oscilloscope. If the

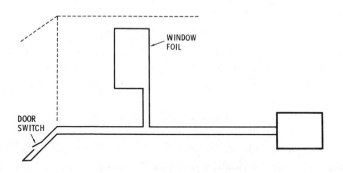

Figure 17-13.
Unequal voltages are induced in unsymmetrical conductors.

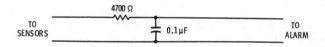

Figure 17-14.
Using a low-pass filter to slow down the response of a system.

circuit is electrically symmetrical there should be very little of anything displayed on the oscilloscope. If there is some dissymmetry, there will usually be a substantial amount of 60 Hz displayed. Additional information can be gained by rapidly turning a light on and off while watching the scope. If any transients are seen, the wiring is susceptible to pickup.

The type of corrective action that can be taken with the wiring from a sensor of a system depends on the type of system. The easiest one to work on is the electromechanical system. This system only responds to a change in dc level produced by the opening of a switch. Such a system can often be cleaned up by installing a low pass filter in series with the wires from the sensing switches. Fig. 17-14 shows an RC filter connected in series with one of the circuits described in an earlier chapter. Here, the filter will effectively slow down the response of the circuit. Inasmuch as only very large transients will affect an electromechanical alarm, the transient will often be gone before the system has a chance to respond.

In audio systems, the filtering is more difficult. Often a filter that will suppress transients will kill the audio signal that is intended to trip the alarm.

THE OUTPUT WIRING

There is a tendency to think of any inteference as entering any system through its input circuits. In alarm systems this type of entry is common. However, it is also possible for surges to get into a system through the output wiring—the wiring that connects the system to the alarm device.

Fig. 17-15 shows an example. Here the leads from a siren are connected to the SCR in the trigger circuit. If a stray surge is picked up by these leads and happens to have the proper polarity, the SCR might be turned on by the dV/dt effect described earlier. A snubber circuit will minimize these effects, but the use of a relay in the SCR anode circuit will provide even greater isolation.

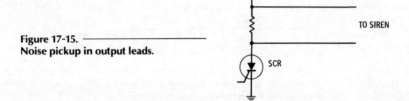

Figure 17-15.
Noise pickup in output leads.

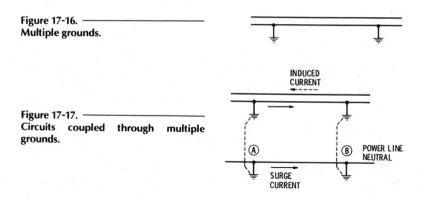

Figure 17-16.
Multiple grounds.

Figure 17-17.
Circuits coupled through multiple grounds.

INDUCED CURRENT

POWER LINE NEUTRAL

SURGE CURRENT

The important thing to remember is that the output wiring, while not as likely to cause problems as the input wiring, can be a trouble spot.

THE GROUND LOOP

A term that is used frequently by electronics technicians, but one that is not very well understood, is "ground loop." Whenever a problem of hum or noise is solved by changing the grounding arrangement of a system, the trouble is explained by simply stating that a "ground loop" existed prior to the change. Often, what is attributed to a ground loop is associated with multiple grounds, but just why multiple grounds should cause problems isn't particularly clear.

Fig. 17-16 shows the ground wire of a system that is grounded at both ends. Looking at this figure we can't see any reason why it should be troublesome. In fact, what causes the problems is what *isn't* shown in Fig. 17-16. The ground symbol which is used in Fig. 17-16 can mean many things. In the ideal case it represents a connection to the earth. It can also represent a connection to anything that is connected to the earth such as a water pipe or a grounded structure. What isn't shown is what else might be connected to ground at approximately the same locations.

For example, consider the arrangement of Fig. 17-17. Here we show our same ground wire, but we also show the ground wire of another system. It may be the ground of a motor circuit, or the neutral of a power line. No matter what it is, it is easy to see that it is effectively connected in parallel with our ground wire.

Whenever any current such as a surge flows between points A and B in Fig. 17-17, the current will divide among the parallel conductors. In this case they are the neutral wire, the ground itself, and our ground wire.

Even this doesn't tell the whole story. We have shown that there might be currents from other devices flowing through whatever we use as a ground in a system. We also know that this current may find its way into the ground wire of our alarm system. Now we see that the ground wire is in close proximity to the other wire of our system. When a surge current, such as that shown in the figure flows in one of two closely coupled conductors, another current will be induced in the other conductor. Now we have an arrangement that can clearly cause trouble. It is easy to see that the ground current and the induced current together can cause a significant voltage across the input terminals of the alarm system. If the ground current happens to be a large surge, a false alarm may well result.

The example that we have given is really quite simplified. Usually there are many different things that can cause substantial currents in anything that we might use for a ground. Thus the first step in eliminating ground loops is to be sure that any grounded wire that we use in a system is grounded at one point only. In many cases this will clear up the trouble.

Unfortunately, in instances where it is necessary to run long wires between two components of a system we do not always have complete control of how the various elements of a system are connected to a common ground. For example, it might be necessary to run a rather long lead between a home-constructed alarm system and a bell or siren that is a purchased item. The siren might well have its own grounding arrangement to the neutral of the power line. A ground loop in a system like this might be difficult to trace and even more difficult to eliminate. Often the problem is solved by simply experimenting with different ground connections. Another approach is to use separate conductors for the circuit ground and the shield ground.

At any rate, when surges find their way into a system, the presence of a ground loop is to be suspected.

INDEX